'The Supreme Triumph of the Surgeon's Art'

A Narrative History of Endocrine Surgery

Perspectives in Medical Humanities

Perspectives in Medical Humanities publishes scholarship produced or reviewed under the auspices of the University of California Medical Humanities Consortium, a multi-campus collaborative of faculty, students and trainees in the humanities, medicine, and health sciences. Our series invites scholars from the humanities and health care professions to share narratives and analysis on health, healing, and the contexts of our beliefs and practices that impact biomedical inquiry.

General Editor

Brian Dolan, PhD, Professor of Social Medicine and Medical Humanities, University of California, San Francisco (UCSF)

Recent Titles

Paths to Innovation: Discovering Recombinant DNA, Oncogenes and Prions, In One Medical School, Over One Decade
By Henry Bourne (Autumn 2011)

Clowns and Jokers Can Heal Us: Comedy and Medicine
By Albert Howard Carter III (Autumn 2011)

The Remarkables: Endocrine Abnormalities in Art
By Carol Clark and Orlo Clark (Autumn 2011)

Health Citizenship: Essays in Social Medicine and Biomedical Politics
By Dorothy Porter (Spring 2012)

What to Read on Love, not Sex: Freud, Fiction, and the Articulation of Truth in Modern Psychological Science
By Edison Miyawaki (Fall 2012)

www.medicalhumanities.ucsf.edu | brian.dolan@ucsf.edu

This series is made possible by the generous support of the Dean of the School of Medicine at UCSF, the Center for Humanities and Health Sciences at UCSF, and a Multicampus Center Research Program grant from the University of California Office of the President.

'The Supreme Triumph of the Surgeon's Art'

A Narrative History of Endocrine Surgery

Edited by
Martha A. Zeiger, MD
Wen T. Shen, MD MA
Erin A. Felger, MD

First published in 2013
by the UNIVERSITY OF CALIFORNIA MEDICAL HUMANITIES PRESS.
BERKELEY — LOS ANGELES — LONDON

© 2013
University of California
Medical Humanities Consortium
3333 California Street, Suite 485
San Francisco, CA 94143-0850

Cover Art George Stubbs, Rhinoceros, c. 1790/91.
From *The Rhinoceros from Durer to Stubbs 1515 - 1799,* Vol. I
(London: Philip Wilson Publishers Ltd., 1986)

Editors' photograph by Ronald Weigel

Library of Congress Control Number: 2013932074

ISBN 978-0-9834639-8-6

Printed in USA

Contents

Acknowledgements

The editors would like to thank those who have made this book possible:

First and foremost, Brian Dolan, for his tireless support and dedication to medical humanities publishing.

Catherine Jones and Joy Oson, who provided invaluable assistance during all phases of this project.

Our colleagues and friends in endocrine surgery, many of whom provided chapters for the book.

Our spouses John Britton, Martha Nicholson, and Ripley Rawlings; and our children Tenaya, Zachary, Evan, Elena, Isabelle, and Jack.

Above all, heartfelt thanks to our mentors and patients. This book is dedicated to them.

Contributors

Peter Angelos, MD, PhD
Professor of Surgery
University of Chicago

Orlo H. Clark, MD
Professor of Surgery
University of California, San Francisco

Peter F. Crookes, MD
Associate Professor of Surgery
University of Southern California

Karen M. Devon, MD
Endocrine Surgery Fellow
University of Chicago

Erin A. Felger, MD
Assistant Professor of Clinical Surgery
Georgetown University

Walter B. Goldfarb, MD
Chief of General Surgery (retired)
Maine Medical Center
Clinical Professor of Surgery
Tufts University

Roger J. Grekin, MD
Professor of Internal Medicine
University of Michigan

Raymon H. Grogan, MD
Assistant Professor of Surgery
University of Chicago

Shelby A. Holt, MD
Associate Professor of Surgery
University of Texas, Southwestern

George L. Irvin III, MD
Professor Emeritus of Surgery
University of Miami

Edwin L. Kaplan, MD
Professor of Surgery
University of Chicago

Sharone Kaplan, BA, MSW
Data Coordinator
Department of Medicine
University of Chicago

Geoffrey W. Krampitz, MD
Resident, General Surgery
Stanford University

Kathleen C. Lee, MD
Research Fellow
Johns Hopkins University School of
Medicine

Bruno Niederle, MD
Professor of Surgery
University of Vienna

Jeffrey A. Norton, MD
Professor of Surgery
Stanford University

Wen T. Shen, MD, MA
Assistant Professor of Surgery
University of California, San Francisco

Norman W. Thompson, MD
Professor of Surgery Emeritus
University of Michigan

Jon A. van Heerden, MB, ChB, MS
Professor of Surgery Emeritus
Mayo Clinic
Adjunct Professor
Medical University of South Carolina

Stuart D. Wilson, MD
Professor of Surgery
Medical College of Wisconsin

Martha A. Zeiger, MD
Professor of Surgery
Johns Hopkins University School of Medicine

Foreword

The history of endocrine surgery is rich with stories that have been passed down from generation to generation of surgeons. They remain a powerful teaching tool for medical students, surgical residents, and endocrine surgery fellows today. While these stories are well-known to surgeons worldwide, numerous changes in the facts and details have accumulated during repeated renditions throughout the years. Furthermore, no volume containing the definitive versions of these stories exists. We thus sought to capture and share the signature stories of the history of endocrine surgery, as retold by current leaders in the field.

Our book contains 16 narrative and pictorial accounts of these seminal stories, in roughly chronologic order. The narratives span centuries and cross continents, from Galen's descriptions of the recurrent laryngeal nerve in ancient Rome to modern-day developments in molecular genetics, radiographic imaging and intraoperative hormone measurements. We hope they will provide a historical context and colorful backdrop to the current practice of endocrine surgery, and serve as the foundation for future advancements in the field.

Martha A. Zeiger, Wen T. Shen, and Erin A. Felger

Pain, Suffocation, Sanguination, Suppuration, and Hysteria

A Short History of Problems Associated with Early Thyroid Surgery with Particular Reference to the Recurrent Laryngeal Nerve

Edwin L. Kaplan
Karen M. Devon
Raymon H. Grogan
Peter Angelos

> For thousands of years, probably, goiter has been a familiar malady. An unsightly and frequently fatal disease, it was accepted as an inoperable affliction or dispensation of Providence in communities where it prevailed and paraded the streets exciting the curiosity of the populace in towns where it was unusual. The sufferers sought relief from suffocation, difficulty in swallowing, failure of the heart, and from a distressing disfigurement. Thus, this conspicuous tumor of the neck was a perpetual challenge to the physician, and to the surgeon a stigma...
>
> William Halsted, 1920[1]

Goiters have probably existed around the world since the beginning of mankind, especially in areas with low iodine content in the soil or water. The early Romans called this condition a "bronchocele." In Italian it is a "gozzo," in French a "goitre," and in Spanish a "bocio." In Western culture goiters were commonly described in the mountainous regions of Northern Italy, Switzerland, Austria, France, and Germany. In the United States, the Midwest was a goiter belt and the Mayo Clinic and Cleveland Clinic came to prominence largely because of their expertise in treating the vast numbers of patients with goiters in those areas.

Goiters have been commonly depicted in artwork over the ages. Michelangelo drew himself with a goiter and complained that he had developed this infirmity after painting the ceiling of the Sistine Chapel.[2] He attributed it to the cramped position that was necessary for him to complete this work. He wrote in Sonnet V:

Already I have a goiter from this toil such as water gives the cats of Lombardy.[2]

Goiters and their complications continue to be prevalent around the world despite efforts to treat large populations with iodine therapy.[3] Cretinous children who experienced growth retardation, neuromuscular problems, deaf-mutism, and mental retardation were very common in Europe in the 1800s and early 1900s. It is estimated that even today 50 to 100 million children around the world may suffer from consequences of iodine deficiency. Only 150-300 mcg of iodine is needed daily, at a cost of 4 cents per day per person, to prevent this dreaded disease.

Why were goiters so challenging and a stigma for physicians, especially surgeons, for so many years? Undoubtedly, this was because treatment for them was limited and ineffective. Surgical treatment in particular was not only limited but was often deadly in the early years when it was attempted.

Seaweed and sponges were used as treatment for goiter for thousands of years, perhaps first in China. These substances could be taken orally, rubbed onto the skin, or even injected directly into the goiter. Iodine was discovered in burned seaweed in 1811 by Bernard Courtois of Paris[4] and was soon used by Coindet in 1820[5] with some encouraging results. Some smaller goiters shrank and even disappeared. However, in most severe cases of large and long-standing goiters, iodine therapy was ineffective. Furthermore, in large doses, iodine was found to be toxic. Sometimes this treatment even resulted in thyrotoxicosis. Its use was therefore limited and even potentially harmful. In many cases, surgery was the only option. Patients with large goiters frequently developed trouble breathing and swallowing. In extreme cases they could suffocate!

Early attempts by surgeons to reduce the size of goiters involved non-cutting techniques such as cauterization of the mass, or placement of a single or multiple setons in order to necrose the growth.[6] Early records of setons used for this purpose go back to Roger Frugardi of the School of Salerno in the twelfth century. Dupuytren in 1833 described a goiter the size of two fists that "lost two-thirds of its size" in 17 days.[7] Setons were not used after approximately 1860 because deaths from hemorrhage, inflammation, and even air embolism were reported.[8]

Other rarely-used methods to decrease the size of goiters involved attempting to drain cysts of the thyroid, ligation of an individual artery or arteries feeding the goiter, and tying off a pedunculated nodule with a shoestring. Finally, a few brave surgeons attempted enucleation of small thyroid masses.

These early attempts at thyroidectomy were treacherous and brutal by our modern standards. No general anesthesia was available until the 1840s. Pain and suffering were pervasive; therefore, speed was of the essence. While the times of early amputations were measured in seconds, some early thyroid operations lasted only several minutes. The first successful thyroidectomy using general anesthesia in which ether was used was performed by Nikolai Pirogoff in St. Petersberg in 1849.[9] Incisions were usually made vertically in order to avoid the large vessels of the neck. Hemostasis was a universal problem because instruments for hemostasis were at best crude and at a premium, and were limited in design and number. The first effective hemostatic forceps were introduced by Wells and Pear in 1874.[10] Techniques of ligation of vessels and even suture materials were in their early stages of development. Asepsis was unknown. Semmelweis was famously vilified in Vienna for even suggesting in 1847 that a surgeon should wash his hands after coming out of the autopsy room and before delivering babies in the hospital.[11] One wonders if instruments that were used were even washed; sterilization of instruments began much later, in 1886. Antiseptics in medicine were not introduced until 1867 by Lister in Glasgow when he sprayed carbolic acid onto the surgical field, first in open fractures.[12] Intraoperative antisepsis with a cap and gown was introduced by Gustav Neuber in 1883. Septicemia leading to death was common. The concept that pus was "laudable," first promoted by Galen, was still in vogue. Healing of wounds by closing the skin, so called primary intention, was uncommonly practiced. Instead, wounds were commonly left open to heal gradually. Astringent or caustic substances were commonly packed into the wound to promote healing, or in the case of the thyroid, to get rid of "the follicle," the name given to the remaining part of the goiter. Thus, healing, when and if it occurred, was a long, tedious, dangerous, and painful ordeal.

William Halsted, in "The Operative Story of Goiter" could trace only eight documented cases of thyroid operations in which a scalpel was definitely used from 1596 to 1800, and 69 others between then and 1850.[2] The mortality of these operations was 41%. Thus, thyroidectomy was condemned and banned by the French Academy of Medicine in 1850 because of the high death rate and the lack of technical improvements.[13] Samuel D. Gross, Professor of Surgery at Jefferson Medical College in Philadelphia, felt strongly that operating on the thyroid gland was too hazardous, and in 1866 in his "System of Surgery" he wrote that, "no honest or sensible surgeon…would ever engage in it."[9]

This was the state of affairs when Theodor Billroth began operating on thyroid masses first in Zurich and later in Vienna. Of his first 36 thyroidectomies, 16 died, mostly of sepsis. For a period of time Billroth mostly abandoned thyroidectomies except to treat asphyxia. He did not resume thyroidectomy until 1877, two years after he had been using Lister's antisepsis. By 1881 he had performed 48 operations with an eight percent mortality rate.[1]

Theodor Kocher, who at age 31 became Professor of Surgery at Berne, Switzerland in 1872, did more to advance thyroid surgery than any other surgeon. A master surgeon, using careful operative techniques, he reduced the mortality for goiter to less than one percent by the beginning of the twentieth century.[1]

Two other problems became apparent during this period. The first, reported extensively from reviews of Billroth's patients, was tetany.[2, 8, 14] No one knew what caused this malady, but some patients died of this disorder. The parathyroid glands were not discovered in man until 1880. Their relationship to tetany was at first unknown. Surgeons proposed that patients with tetany had vascular or nervous problems. Some thought that the female patients who developed tetany were hysterical. It was not clarified that tetany was definitely due to hypocalcemia from damage to the parathyroids until the early 1920s. Before that time many physicians and surgeons proposed that the function of the parathyroid glands was to detoxify unknown substances.

The second problem was postoperative hypothyroidism and cretinism. In 1867, Paul Sick of Stuttgart described a 10-year old boy who had a total thyroidectomy.[8] He had been "joyous and lively" and 10 months later became quiet and dull. The boy lived 18 years thereafter as a cretinous dwarf.

In 1874, Kocher performed a total thyroidectomy on Marie Bichsel, aged 11.[15, 17] She was described as a "spirited and joyous creature" but later became "peevish and dull." He heard from her physician that she had become cretinoid. Kocher called back his patients and found that most of those who had had total ablation showed the same features. He called the condition "cachexia strumipriva." Kocher and Reverdin decided to stop performing total thyroidectomies.[16] Kocher thought that this disorder was caused by an injury to the trachea resulting in chronic asphyxia. By the end of the nineteenth century, there was clear evidence that cachexia strumipriva was similar to myxedema, that myxedema was caused by the absence of the thyroid, and that replacement with thyroid hormone extract reversed this disease. This was the first evidence that a substance from the thyroid was physiologically active. Kocher received a Nobel Prize for his work on the physiology, pathology, and surgery of the thyroid in 1909.

Billroth's patients had considerable problems with tetany, while Kocher rarely saw this disorder, but described cachexia strumipriva, not often recognized by Billroth. Halsted hypothesized that the difference was that Kocher, operating carefully in a relatively bloodless field, removed only the thyroid completely.[1] Billroth, operating more rapidly, clamping tissues "en masse" with less concern about hemostasis, could more readily leave thyroid tissue behind and remove the parathyroids or damage their blood supply. Perhaps referring to Billroth, Kocher wrote:

> Surgeons who take risks and operate by the clock are exciting from the onlooker's standpoint, but they are not necessarily those in whose hands you would by preference choose to place yourself.[17]

The last problem, injury to the recurrent laryngeal nerve or nerves, plagued surgeons for thousands of years. Dysphonia after operation was thought by some to be due to hysteria. The recurrent laryngeal nerves, discovered over 2000 years before, became the ultimate problem following thyroidectomy. The following case study from the early 1700s illustrates many of the obstacles that we have discussed.

> Ursula Curti of Cauriago, a young woman of eighteen years, would have been most graceful and beautiful, gifted of extraordinary spirit, and one of the greatest of her circle, had she not been deformed by a goiter the size of the head of a small boy, so large that it rose above her chin by a width of three fingers. Desirous of freeing herself from such a dishonorable deformity, she had visited many physicians, surgeons, and general practitioners, all of whom tried in many, different ways, and with various cures both internal and external to extirpate, but all were unfruitful. But finally, advised by Father Maestro Ciardi Domenicano, came to me here in Scandiano in the year 1717. I saw the patient, but, frightened by the unusual size of this tumor, declined the undertaking....I decided finally to 'put my hands in the pasta'... and attempt an extirpation with every caution requisite to such an arduous and dangerous operation.

Thus wrote Dr. Fulvio Gherli, Doctor of Philosophy and Medicine, who practiced in Scandiano, Italy, in 1724.[18]

Dr. Gherli made a vertical incision in the middle of "this nasty tumor" with a sickled knife and "two pounds of nasty liquid poured forth from the tumor." He applied swabs and tampons soaked in well beaten egg batter and astringent powders. After several weeks he changed to corrosives--lint soaked in precipitate, sublimate, alum and vitriol mixed with butter in order to eradicate "the follicle." After 15 days of the later treatment, in order to operate more adroitly, "I decided to dilate the goiter."

Several hours after enlarging the wound vertically with a knife, he found the patient "completely and thoroughly soaked in blood down to her feet." He packed the wound with unspun cotton soaked in styptic water and the bleeding stopped. Eventually the wound healed and she was cured. Halsted refers to this report as one of the first cases of thyroid surgery that he could document in the written literature.[1]

When discussing this case, Dr. Gherli wrote:

> The accident of hemorrhage is a minor evil....While there are others more terrible and frightening, the cutting of the recurrent nerves is dangerous in the highest degree, [for when] this unfortunately occurs, either the patient dies of it miserably or at least loses for the rest of his life the most beautiful prerogative given to man by God, which is (la favella) speech; *but this danger can easily be avoided by that Surgeon who, with the provision of Anatomy, knows the site of these nerves.*[18]

How is it possible that so much was known about the recurrent laryngeal nerve that in 1724, every well-trained surgeon would know the anatomical position of these nerves, the complications associated with them, and how to avoid damaging them at operation? We examined the history of the recurrent laryngeal nerves, which extends back over 2000 years.

The first reference found of the control of voice is in the Sushruta Samhita, written in India in the sixth century B.C.[19] Sushruta was the founder of the Ayurveda system of medicine which still exists to this day. In his *Samhita*, one reads,

> There are four Dhamanis (arteries) about the two sides of the Kantha-Nadi (wind pipe). Two of them are known as Nila (meaning glistening-white in

Sanscrit), and the other two as Manya (meaning purple). One Nila and one Manya are situated on either side of the larynx. An injury to any of them produces dumbness, and a change of voice (hoarseness) and taste.[19] An important Marma, or critical area, number 16, occurs at the angle of the jaw.

Thus, in 600 B.C., injuries to the vessels of the neck were thought to cause hoarseness.

In 100 A.D., Rufus the Ephesian noted that it was the nerves and not the vessels that were responsible for the faculty of voice. He wrote,

> The ancients applied the name carotid (drowsy or stupor) to the vessels that cross the neck because their compression produces drowsiness and aphonia. Now, however, we realize that these symptoms result from compression of the nerves and not the vessels.[20]

However, it was Galen who first described the recurrent laryngeal nerves in detail in the second century A.D. Galen, early in life, was surgeon to the gladiators in Pergamum and described different neurologic conditions resulting from their injuries. While Greek philosophy and medicine held the heart to be most important--the seat of intellect and learning--Galen recognized the importance of the brain. He was delighted when he found a nerve on each side of the neck which went down toward the heart and then reversed course and reascended to the larynx. Nerves were thought to contract along with tendons and muscles. In order to contract the laryngeal muscles, the pull had to be from below, he thought, and here was just such a nerve which came from the brain.

> In the passage of the nerves across the thorax, a branch reascends on each side by the same pathway which it took before in descending; thus it accomplishes a double course...It reascends from there to the larynx where the nerves insert themselves into the muscles in question.[21]

> I call these two nerves the recurrent nerves (or reversivi) and those that come upward and backward on account of a special characteristic of theirs which is not shared by any of the other nerves that descend from the brain.

He took great pride in this discovery.

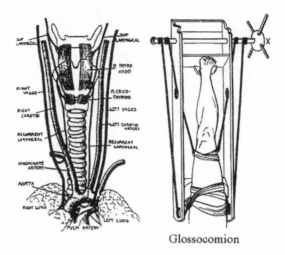

Glossocomion

Figure 1: Galen wrote that the recurrent laryngeal nerves gained strength to close the vocal cords because of their unique design of a pulley system (left). This mechanism was similar to a glossocomion, a common instrument of his day which utilized a pulley system to reduce fractures (right). From "Oeuvres D'Oribase," Paris, 1862, volume 4.

All these wonderful things, which have now become common property, I was the first of all to discover, no anatomist before me ever saw one of these nerves, and so all of them before me missed the mark in their anatomical description of the larynx.[22]

He felt that the nerve or cord gained great mechanical advantage by a pulley action, similar to a *glossocomion* which was a popular device of that day for reducing fractures (Figure 1).[22, 23] He dissected these nerves in many animals, even in swans, cranes, and ostriches because of their long necks and marveled at the mechanical advantage created by the pulley system that enabled such small nerves to open and close the muscles of the larynx.

Galen recognized in studies on the living pig that "if one compresses the nerve with the fingers or a ligature" or if one cuts the nerve, the pig stopped squealing and the muscles of the larynx on that side ceased to work.[24]

He gathered the elders of Rome and in order to impress them with his greatness and knowledge, operated on the neck of a live squealing pig. When he cut the recurrent laryngeal nerve, the pig stopped squealing. His dissection on the living pig is depicted in a beautiful medieval illustration (Figure 2).[24]

Galen described two children who were operated upon by surgeons ignorant of anatomy:

So when someone cut out from (the slave's) throat the swollen glands which were deeply set, and did not cut out the membranes with his instrument, but instead tore them out with his nails, he unwillingly, due to his ignorance, removed the surrounding nerves at the same time; and in this manner he freed the slave of his swollen glands, but rendered him mute.

Similarly, someone else while performing an operation on another child rendered him half mute, evidently having damaged only one of the nerves.... Everybody found it strange that the voice was damaged, although the larynx and trachea remained intact. But when I demonstrated to them the phonetic nerve (i.e., the recurrent nerve), their astonishment abated.[25]

Because of Galen's fame and the spread of his teachings, the recurrent laryngeal nerve was discussed by many surgeons and anatomists thereafter. Aetius, in the sixth century, wrote "In the case of the throat glands, the vocal nerves must be carefully avoided...[otherwise] the patient is bereft of his voice."[26] Paulus Aegineta, in the seventh century, again stressed that when operating in the neck, "avoiding in particular the carotid arteries and recurrent nerves" must be exercised.[27]

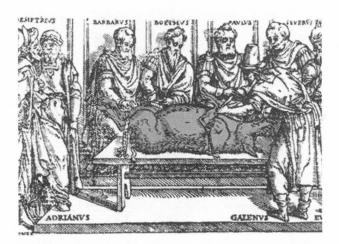

Figure 2: Galen demonstrating the recurrent laryngeal nerve to the elders of Rome in the living pig. When the nerve was divided, the pig's squealing ceased and it became mute. From Galeni Librorum Quarte Classis. Venetijs Apud Iuntas, 1586.

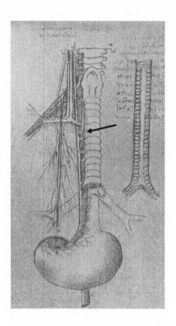

Figure 3: Leonardo Da Vinci in 1503 drew the course of the right vagus nerve in an animal. This illustration might be the first to demonstrate the recurrent laryngeal nerve (arrow).

Arabic medical literature of the ninth through twelfth centuries also contains references to the recurrent laryngeal nerve. Many of the ancient manuscripts, particularly those of the Greeks, were collected and translated by the Arabs during that period. Abul Kasim (Albucasis, 1000 A.D.) is credited with the first recorded description of a thyroidectomy. He echoed the same warnings with regard to the recurrent laryngeal nerve: "Be most careful not to cut a blood-vessel or nerve."[28] He also described a slave girl who stabbed herself in the neck. The artery and vein had not been cut, but she developed hoarseness.

During the Middle Ages, the experiments on the recurrent laryngeal nerves in pigs were repeated in the Salernitan demonstrations. Just as Galen had done, a squealing pig was placed on its back and when the recurrent nerve was divided, the pig stopped squealing. "Although some animals, such as monkeys, are found to resemble ourselves in external form, there are none so like us internally as the pig."[28] Then, the dissection of the motivi (vagus nerves) and reversivi or recurrent laryngeal nerves is described. Salerno, in southern Italy, is of historical interest because of its medical school—the earliest in Europe.

Progression of the anatomic knowledge of the recurrent laryngeal nerve was demonstrated during the Renaissance. In 1503, Leonardo Da Vinci drew what may be the first anatomical representation of the recurrent laryngeal

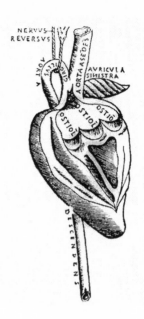

Figure 4: Jacopo Berengario Da Carpi clearly demonstrated a representation of the left recurrent laryngeal nerve in this drawing of the heart. Bologna, 1523.

nerve, possibly in an ape (Figure 3).[30] It should also be noted that the first drawings of the thyroid gland are attributed to Da Vinci. Berengarius' drawing (1523 A.D.) of the heart clearly demonstrates a left recurrent laryngeal nerve (Figure 4). He warned that if a surgeon is ignorant of anatomy, "half-impaired" or "fully-impaired" voice may result.[31]

Vesalius, writing in 1543, was particularly interested in the recurrent laryngeal nerve since "nothing is more delightful to contemplate than this great miracle of nature," he wrote. His drawing of Cupids operating on the neck of the pig are reminiscent of Galen's former operations (Figure 5a). He also produced excellent anatomical drawings of the recurrent laryngeal nerves (Figures 5b and 5c).[32]

Images of beautiful anatomical dissections by Charles Estienne in Paris from 1545 and 1546 clearly demonstrate both recurrent nerves (Figure 6).[33] Finally, Hieronymus Fabricius of Aquapendente, from Padua, in 1600 (Figure 7) and Julius Casserius in 1601 (Figure 8), published very detailed illustrations of the larynx and recurrent laryngeal nerves.[34, 35]

Hence, by 1717 when Fulvio Gherli operated on the beautiful young girl described in the beginning of this chapter, a great deal was known not only about the anatomy of the recurrent laryngeal nerves, but also the complications when one or both of them were cut or damaged, and ways to avoid these problems.

Vesalius, in his masterpiece De Humani Corporis Fabrica 1543, illustrated the recurrent laryngeal nerve and stated that "nothing is more delightful to contemplate than this miracle of nature."

Figure 5a: Cupids operating on the neck of a living pig, as Galen had done in the past.

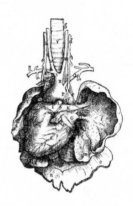

Figure 5b: Anatomy of the neck and chest, demonstrating the left and right recurrent laryngeal nerves.

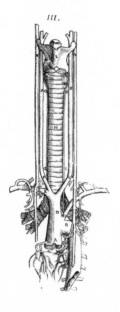

Figure 5c: The course of the recurrent laryngeal nerves.

Figure 6: Taken from the beautiful anatomical dissection of Charles Estienne,
1546. The recurrent laryngeal nerves
are shown.

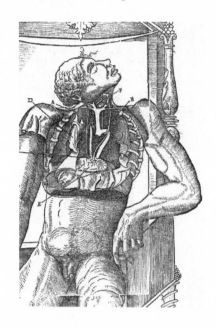

But surgeons still had troubles. Billroth's group, reported by Wolfler in 1882, and discussed earlier in this chapter, had a 40% mortality for thyroidectomy for goiter while he was in Zurich prior to 1867 and an 8.3% mortality for goiter in the antiseptic period, from 1877 to 1881.[36] Five patients (10.5%) required tracheostomy. Unilateral nerve injuries were reported in 25% (11 of 44) and bilateral nerve injuries in 4.5% of the later group (1877 to 1881).

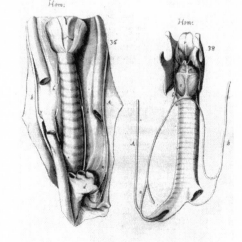

Figure 7: The anatomical dissection of Hieronymus Fabricius of
Aquapendente, Padua, 1600, of
the trachea, larynx, and recurrent
laryngeal nerves.

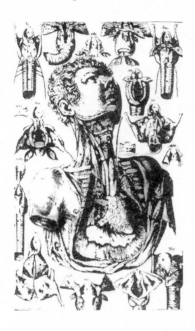

Figure 8: The larynx and recurrent laryngeal nerves in the dissections of Julius Casserius, Padua, 1601.

Jankowski reported a 14% incidence (87 of 620 patients) of recurrent nerve injuries during goiter operations prior to 1885.[37] Halsted later commented that undoubtedly the number was greater since laryngeal exams were not routine.[38]

It was Theodor Kocher of Berne who brought his operative mortality of thyroidectomy from 14.8% in 1882 to an eventual level of less than 0.18% in 1898.[39] His meticulous technique resulted in an incidence of recurrent nerve injury similar to that of surgeons today.

Many additional advances in the anatomy and physiology of the laryngeal nerves and of the muscles of the larynx and of the vocal cords have been made. These are discussed in an excellent monograph by William H. Rustad.[40]

However, a critical chapter of the recurrent laryngeal nerve was played out in the United States. Dr. George Crile, in his textbook "Diagnosis and Treatment of Diseases of the Thyroid Gland" in 1932 wrote,

> Every student of surgery knows the general position of the recurrent nerves, and yet the greatest tragedies which follow thyroidectomies pertain to these structures. Even the surgeon who has had much experience in operations on the thyroid gland reviews the position of the recurrent nerves as an evil memory. The hazard is not due, however, to the anatomical location of these nerves, but rather to the vulnerability of their structures, to the neighbor-

Figure 9: Routine dissection
and demonstration of the
course of the recurrent
laryngeal nerve during
subtotal thyroidectomy by
Dr. Frank Lahey, 1938.

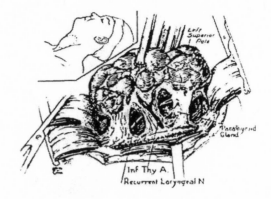

hood of fixation, to adhesions, and to certain characteristics of the nerve
conduction.

As compared with peripheral nerves, the recurrent nerves are exceedingly
soft...and the slightest direct or even indirect pressure on the recurrent
nerve interferes with nerve conduction....And it is this extreme vulnerability
that is the first and the most important factor in the production of abductor
paralysis.

It is certain that if the nerve trunk is directly exposed in the course of the
operation, the exposed nerve will be covered with scar formation. Scar tissue
is capable of producing a block in the action current, hence, causing a physiologic severance of the nerve.[41]

Crile recommended leaving the posterior capsule of the thyroid in each thyroid resection. "The area near the nerve is 'no man's land.' It is not to be pal-
pated; it is subjected to the least possible traction and no division of tissue is
made. By these precautions temporary and permanent injury of the recurrent
laryngeal nerve may be completely eliminated...."

This philosophy of never looking for the nerve, which influenced an entire generation of surgeons and is still taught and practiced by some, may actu-

ally result in higher rates of permanent nerve injuries, in our opinion.

Later, in 1938, Lahey reported on over 3,000 thyroidectomies performed by his fellows and staff over a three-year period.[42] (Figure 9) The recurrent laryngeal nerve was dissected in virtually every case. Careful dissection would "not increase but definitely decrease the number of injuries to the recurrent laryngeal nerves," he wrote. Lahey's work with its emphasis on anatomy set the course and direction for modern thyroid surgery.

In conclusion, because of its many pitfalls and serious associated complications, thyroidectomy has been a perpetual challenge and a stigma to surgeons over several millennia. Halsted wrote,

> is there any operative problem profounded as long ago and attacked by so many which has cost so much thought and endeavor and so many lives before its ultimate solution was achieved?[43]

Furthermore, Halsted, in 1920, called thyroidectomy "the supreme triumph of the surgeon's art...a feat which can be accomplished by any really competent operator without danger of mishap..." However, serious complications still occur. Even today, surgeons must be mindful of the failures of the past and operate with great care and expertise in an unhurried manner if they are to attain the best possible results.

Acknowledgments

We thank Mrs. Patricia Schaddelee for her excellent help preparing this chapter. A portion of this chapter was previously published in *World Journal of Surgery* (2009 Mar;33(3):386-93). Reprinted with permission.

References

1   Halsted WS. The operative story of goiter. The author's observations. The Johns Hopkins Hospital Report, 1920; 19:71-257.
2   Merke F. History and Iconography of Endemic Goiter and Cretinism. Hans Huber Publishers, Berne, 1984; p. 155.
3   Eastman CJ, Zimmerman MB. The Iodine Deficiency Disorders. Chapter 20, Thyroidmanager.org.
4   Abraham GE. The history of iodine in medicine. Part 1. From discovery

to essentiality. The Original Internist 2006; 13:29-36.

5   Coindet JF. Decouverte d'un nouveau remede contre la goiter. Ann Clin Phys 1820; 15:49.

6   Ericksen JE. The Science and Art of Surgery. London, Longmans, Green. 1872; 2:330-333.

7   Dupuytren G. Treatment of goiter by the seton. Lancet 1833; 2:685-686.

8   Wellbourn RB. The History of Endocrine Surgery. Chapter 2. The Thyroid. Praeger, New York, 1990; p. 25.

9   Gross SD. A System of surgery. Vol. II, 4th Ed., H.C. Lea, Philadelphia, 1886.

10   Harvey SC. The History of Hemostasis. Hoeber, New York, 1929.

11   Nuland SB. The Doctors' Plague: Germs, Childhood Fever, and the Strange Story of Ignac Semmelweis. WW Norton, 2003.

12   Lister J. The Collected Papers of Joseph Baron Lister. Oxford, Clarendon Press, 1909.

13   Haeger K. The Illustrated History of Surgery. Harold Starke. London, 1988.

14   Sandstrom I. Om en ny kortel bos menniskan och Atskilliga daggdjur Uppsala Lakereforenings Forhandlingar. 1880; 15:441.

15   Kocher T. Ueber Kropfextirpation und ihre Folgen. Arch Klin Chir 1883; 29:254-337.

16   Reverdin J-L. Les accidents consecutifs a l'ablation totale du goitre. Rev Med Suisse Romande 1882; 2:539-540.

17   Zimmerman LM, Veith I. Great Ideas in the History of Surgery. Baltimore. Williams and Wilkins, 1961; 499-518.

18   Gherli F, Osservazione XIII. Gozzo sterminato. Centuria Seconda de Rare Osservazioni di Medicina e Chirugia di Fulvio Gherli, Presso Michele Pigone, Venezia, 1724.

19   The Sushruta Samhita. Volume I. Translated by K.K. Bhighagratna. Calcutta, 1907, p. 185.

20   Oeuvres De Rufus D'Ephese, Daremberg C, Ruelle CE. L'Impremerie Nationale, Paris, 1879.

21   Galen on Anatomical Procedures-The Later Books. Translated by W.L.H. Duckworth. Lyons MC and Towers B, editors. Cambridge University Press, 1962, p. 81.

22   Galen on the Usefulness of the Parts of the Body. Translated by M. Tallmadge May. Cornell University Press, Ithaca, New York, 1968, pp. 364-368.

23 The Doctrine of the Nerves. Spillane JD. Oxford University Press, 1981, p. 21.
24 Galen. Librorum Quarta Classis. Venetijs Apud Iuntas, 1586.
25 Opera Omnia. Clavdii Galeni. D.C. Gottob Kuhn, Editor. Volume VIII. Lipsiae, prostate in Officina Libraria Car. Cnoblochii, 1924, p. 55.
26 Merke F. ibid, p. 88.
27 The Seven Books of Paulus Aeginetus. Translated by F. Adams. Volume II. The Sydenham Society. London, 1866, p. 307.
28 Spink MS, Lewis GL. Albucassis on Surgery and Instruments. University of California Press, Berkley, 1973, p. 332.
29 Corner GW. Anatomical Texts of the Earlier Middle Ages. National Publishing Company, Washington, DC, 1927, p. 51.
30 O'Mally CD, de CM Saunders JB. Leonardo on the Human Body. Dover Publications, New York, 1983, p. 149.
31 Merke, ibid, p. 144
32 de CM Saunders JB, O'Mally CD. The Illustrations from the Works of Andreas Vesalius of Brussels. Dover Publications, New York, 1973, p. 151.
33 Estienne C. De dissectione partuim corporis humani libri tres. Paris, Simon Colinaeus, 1545.
34 Merke, ibid, p. 168.
35 Merke, ibid, p. 170.
36 Wolfler A. Die Kropfexstirpationen an Hofr. Billroth's Klinic von 1877 bis 1881. Wien. Med. Wochenschr. 1882, 32:5.
37 Jankowski F. Lahmungen der Kehlkopfmuskein nach Kropfexstirpation. Deutsche Zeitschr. F. Chir. Leipzig 1885, 12:164.
38 Halsted, ibid, p. 185.
39 Halsted, ibid, p. 174-177.
40 Rustad WH. The recurrent laryngeal nerves in thyroid surgery. Charles C. Thomas, Springfield, 1956, pp. 1-47.
41 Crile G and associates. Diagnosis and Treatment of Diseases of the Thyroid Gland. W.B. Saunders, Philadelphia, 1932, pp. 401-409.
42 Lahey FH. Routine dissection and demonstration recurrent laryngeal nerve in subtotal thyroidectomy. SGO 1938; 66:775-777.
43 Halsted, ibid, p. 71.

2

"Infinite accuracy, intimate care, infinite patience"
Theodor Kocher and the Development of Modern Thyroid Surgery

Orlo H. Clark

The Beginning – A Hypothetical Patient with Goiter

In 1895, a 21 year-old medical student with a neck mass that had gradually enlarged was waiting to see his physician. His neck mass was caused by a goiter that was larger on the right side, measuring 10x9x8cm. The student had noted intermittent dysphagia without change of voice but with mild shortness of breath while exercising. He volunteered that he felt very uncomfortable when he lifted his arms over his head and that his face would turn red when he was in this position. The young man was otherwise in good health and had no family history of any illnesses, including goiter. He lived in Switzerland, which was known to be an iodine-deficient area.

This intelligent student had read extensively about his medical problem and was acutely aware of the current literature regarding goiter and its treatment as well as the morbidity and mortality of thyroid operations. He also knew that Celsus, Galen and Albucasis may have operated on the thyroid glands many centuries earlier but that only recently had removal of the thyroid gland become an accepted form of treatment for enlarged goiters.[1]

Anatomy and Function of the Thyroid Gland

The well-informed student knew that during the sixteenth century the great Renaissance anatomist, Andreas Vesalius of Padua, described a thyroid gland with two separate lobes.[2] He was unaware however, that in about 1500 Leonardo da Vinci had also illustrated the thyroid gland as two separate lobes and regarded it as two glands.[3] Da Vinci wrote "there are glands (the thyroid gland) made to fill up the space where the muscles are missing and to keep the trachea away from the clavicle".[4] DaVinci's drawings were lost for nearly three centuries until they were eventually found in the vast collections of Windsor Castle in England (Figure 1).[3,4] Despite the expertise of the anatomist and artist, however, knowledge of the thyroid gland's function remained a mystery until the nineteenth century.

In the seventeenth and eighteenth centuries several physicians became puzzled about the anatomy and the purpose or function of the thyroid gland. Bartholomeus Eustachius (1537-1637) of Rome was the first to accurately describe a single thyroid gland with the two lobes connected by an isthmus.[3] Eustachius was also the first anatomist to discover the adrenal or suprarenal glands.[3] In 1619 Fabricius ab Aquapendente from Padua documented that goiters arose from the thyroid gland.[3] The British physician Thomas Wharton (1614-73) suggested that the thyroid gland "contributed to the beauty of the throat, particularly in women".[3] In 1656 he was the first to use the term *glandula thyroideus* in his book, *Adenographia*.[3] Although the anatomy of the gland was known by the 1600s, ideas about its function were speculative rather than logical and included the following. The thyroid served a) to moisten the larynx and the trachea, b) to relieve the trachea from pressure, c) to produce an 'inner warmth', d) to regulate the flow of blood to the brain, e) as a blood reservoir for blood returning from the brain, f) to lessen vibrations from the larynx and the trachea, g) to improve the timbre of the voice, h) to round out and beautify the neck, i) to be a blood forming organ, j) to lubricate the larynx or even serve as a "bag of worms".[5,6] Frederik Ruysch (1638-1731), a Dutch anatomist, and subsequently Theophile de Bordeu (1722-1776), a French scientist from the Vitalist school at Montpellier, France were the first to suggest that the thyroid secreted substances into the blood, although they had no convincing objective data to support their opinions.[7,8]

Advances in Surgery and Anesthesia

Surgical treatment of thyroid disorders developed much later than the medical diagnosis of goiter. Surgery in general was both primitive and dangerous until the introduction of general anesthesia in the 1840s by American and European physicians and dentists who documented the benefits of using ether or nitrous oxide during procedures or operations.[9] Crawford Long (1815-1878), Horace Wells (1815 – 1848), Thomas Morton (1819- 1868) and John Collins Warren (1778-1856) demonstrated the effectiveness of the anesthetic agents and Henry J. Bigelow (1816-1890) published about the benefits of using ether in surgery in 1846.[9-11] General or local anesthesia quickly became an essential part of most operations.[12] It resulted in relief of pain for the patients and also

Figure 1: Illustration of the thyroid
gland and larynx by Leonardo da
Vinci (1452-1519)

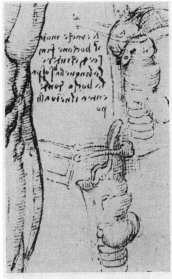

Abb. 46. LEONARDO DA VINCI. Fogli A. Kehlkopfskizzen.

enabled surgeons to operate without rushing, thus making the operating room
a more pleasant and safer place.

Other perioperative problems including sepsis and bleeding, however,
continued to impede the success of surgical procedures. In 1867 Joseph
Lister (1827-1912), Professor of Surgery in Glasgow, Scotland, documented
the benefits of using carbolic acid and hand washing for antiseptic
surgery.[13] In 1883, Lister acknowledged the previous contributions of Ignaz
Semmelweis (1818-1865) and his studies relating to puerperal sepsis.[14] Lister's
use of antiseptic techniques was rapidly adopted in Europe with marked
improvement in operative results.[9] The development of better surgical clamps,
such as homeostatic self-retaining forceps, as reported in 1874 by Spencer
Wells (1818-1897), both enhanced the development of surgical technique by
reducing operative bleeding and contributed to improved patient outcome.[15,16]

These advances in general surgery dramatically improved the results of
thyroid surgery, which was condemned by the French Academy of Medicine in 1850 despite the fact that the French surgeon, Pierre-Joseph Desault
of Paris had performed a successful thyroidectomy in 1781.[17] At this time
prominent surgeons from London, Berlin and Philadelphia adamantly advised
against operations on goiters.[18] This caution was understandable because prior
to 1850 approximately 70 thyroid operations had been done with a mortality
rate of over 40%.[19]

Theodor Billroth

In 1860 at the age of thirty-one, Theodor Billroth (1825-1894) became the chair of surgery at the University of Zurich. While in Zurich and before moving to Berlin, he performed twenty thyroidectomies with a mortality rate of 40%.[20] Seven of his patients died from sepsis and one from hemorrhage. Billroth therefore stopped operating on the thyroid for almost a decade. In 1877 he resumed thyroid surgery, and between 1877 and 1881 he performed 48 thyroidectomies with an operative mortality of only 8.3%.[20] Despite a lower mortality rate, his complication rate continued to be excessive with 8 of his 22 patients who had total thyroidectomies developing tetany and 13 patients having recurrent nerve palsy.[21] Billroth's assistants, Johann von Mikulicz and Anton Wölfler, who in 1886 became chairs of surgery in Krakow and Graz respectively, learned from the complications in their mentor's patients. They recommended subtotal rather than total thyroidectomy to avoid injury to the recurrent laryngeal nerves and parathyroid glands.[22]

Anatomy and Function of the Thyroid and Parathyroid Glands

The parathyroid glands were first identified grossly in an Indian rhinoceros by Sir Richard Owen[22] and in animals and man by the Swedish medical student Ivar Sandström in 1879.[23] Eugene Gley (1857-1930), a biologist from France, was the first to address the function of the parathyroid glands. He reported that thyroidectomy resulted in tetany only when the parathyroid glands were removed.[24] This observation was confirmed by the studies of the Italian pathologists Giulio Vassale (1862-1912) and Francesco Generali (1896-??).[25]

The function of the thyroid gland was still unknown, however, although Caleb H. Parry (1735-1822) in 1825, Robert Graves (1795-1822) in 1835 and Carl Avon Basedow (1799-1854) in 1844 described patients with goiter, exophthalmos and palpitations.[26] In 1859 Moritz Schiff (1823-1896), an anatomist from Geneva, Switzerland, reported that one week after dogs and guinea pigs underwent total thyroidectomy both groups of animals died but the guinea pigs, undergoing the exact same operation, survived a little longer.[27,28] These experiments suggested that the thyroid gland might be essential for life. Some experts, however, questioned whether these animals died as a result of the operation itself or because of removal of the thyroid gland. Few understood the importance of Schiff's findings. Until 1891 the function of the thyroid

Figure 2: Photograph of
Dr. Theodor E. Kocher

remained unknown, and the thyroid was still believed to be a "lesion of the nervous system."[8,29]

Theodor Emil Kocher

What could the Swiss medical student know about the function of thyroid gland in 1895? He was aware of studies by Murray, Fox and Mackenzie who documented that by eating or receiving injected thyroid tissue one could prevent the development of myxedema.[30-32] He was also aware of the accomplishments of the Swiss surgeon Theodor Kocher (1841-1917) (Figure 2). Kocher was born in Switzerland on August 25, 1841 and died on July 27, 1917. He was an excellent student at the University of Berne where he received his medical degree and graduated *summa cum laude*.[33] Following graduation, he spent a year in England with Sir Robert Hutchinson, Sir James Paget and Sir Thomas Spencer Wells. He also studied with Bernard von Langenbeck (1810-1882), George A Lüche (1829-1894) and Theodor Billroth (1829-1894).[20,32,33] From 1866 to 1869 Kocher "was the sole assistant of Lüche at the University of Berne and during this time Kocher introduced (Lister's) antiseptic wound treatment despite the objection of many (colleagues)."[33] In 1872 at age 31, after Lüche left Berne to become Chair in Strasbourg, France, Kocher was named Professor of Surgery and Director of the University Clinic of Surgery at the University of Berne where he remained as Chief of Surgery for 45 years until his death in 1917.[20]

Kocher was aware of the ominous results of thyroidectomy performed by other surgeons. During Kocher's first two years at Berne he performed 13 thyroid resections. He initially followed Billroth's technique with a near vertical incision paralleling the anterior border of the sternocleidomastoid muscle, but subsequently he employed a collar incision that had been introduced by Jules Boeckel, from Strasbourg, in 1880.[34] Six of his patients were short of breath preoperatively, demonstrating the extent of their thyroid goiters. Despite the use of antiseptic techniques, two of these patients died from infection although no one died from hemorrhage.[27] "During the first 10 years of his tenure at Berne, Kocher excised 101 goiters, with a mortality rate of 12.8%"[20] By 1889 his mortality rate in 250 patients with non-toxic goiter was 2.4%, and by 1895 his operative mortality rate for benign goiters was approximately 1%. His superior surgical technique is supported by his report in 1895 that in more than 1,000 patients operated on for goiter only one developed tetany.[35] Kocher reviewed the results of other surgeons in Europe and the USA who had performed thyroid resections from 1850-1877 and reported that the mortality rate in these patients had fallen from 41 to 21%. In 1883 Kocher reported that he had performed 268 thyroid operations since 1877. "The mortality rate for non-malignant goiters had again fallen to 12%, while that for malignant cases was 57%"[34] The reasons for the improved operative results were probably due to better anesthesia, more experience, better surgical instruments and aseptic technique.

When Kocher and several of his peers observed behavioral and physical changes in some patients after total thyroidectomy, the vital function of the gland began to become clear. In 1867 Paul Sick (1836-1900) from Stuttgart, Germany reported that a "joyous and lovely" boy became sluggish and dull following a total thyroidectomy performed by Wilhelm Hahn and Karl Bockshammer.[36] "The boy lived at least 18 years as a cretinous dwarf."[36] In 1882 the Swiss surgeon from Geneva, Jacques-Louis Reverdin (1842-1929), and his cousin Auguste Reverdin (1849-1908) reported several patients "who became feeble and anemic two or three months after he had performed total thyroidectomy."[37] Two patients "developed edema of the face and hands and one looked 'cretinoid'."[37] "Reverdin called this condition 'myxedema operatoire'."[37]

According to Welbourn, Reverdin's results encouraged Kocher to contact his former patient Marie Bichsel on whom he'd performed a total thyroidectomy for a benign goiter nine years earlier in 1874.[37] Kocher observed that after her thyroid was removed, Bichsel had changed from "a spirited and joyous creature" to one who was "peevish and dull," similar to Sick and Reverdins' patients.[37] In contrast, her sister, who had not undergone thyroid surgery, thrived.[37] Examining his previous patients, Kocher found that "30 of his first 100 patients operated on for goiter, whom [he was] able to follow up and reinvestigate, presented a syndrome . . . [he] designated cachexia strumipriva."[39] Twenty-eight of thirty patients who had partial thyroidectomies were well, whereas 16 of 18 who had total thyroidectomies developed myxedema. The clinical changes were more dramatic in children.[40] Kocher presented his findings at the Deutche Gesellschaft für Chirurgie (Congress of German Surgeons) in Berlin in April 1883. He surprisingly attributed the condition of *chachexia strumpiva* to injury to the trachea despite knowing that ten years earlier William Gull (1816-90) a physician from London in 1874 had described a cretinoid condition spontaneously developing in adult life in five women, clearly not related to a tracheal injury.[37] This clinical condition was subsequently named "myxedema" or "mucous edema" by the British physician William Ord (1834-1902), who attributed it to insufficient thyroid secretion.[27] In 1883 Sir Felix Semon (1848-1921), an ENT surgeon, proposed at the Clinical Society of London "that cretinism, myxedema and cachexia strumipriva were all associated with absence or degeneration of the thyroid."[27] The "society (subsequently) appointed a committee with Ord as Chairman to investigate myxedema."[27] They reviewed over 100 reports of myxedematous patients as well as the effects of thyroidectomy in man and animals.[41]

In 1884 Moritz Schiff (1823-96), whose previous experiments on dogs, mentioned earlier in this chapter, again thyroidectomized 60 dogs and reported that animals that had had thyroid tissue previously autotransplanted into the abdomen did not die after thyroidectomy.[27,28,42] In 1884 Sir Victor Horsely, who was also a member of Ord's Committee and a surgeon at University College Hospital, thyroidectomized monkeys and reported first muscular twitchings and convulsions, then myxedema.[43] In 1888 Horsley and Ord's other committee members presented a report at the London Clinical Society in which they agreed with Semon that cretinism, myxedema and cachexia strumipriva were the same condition and were caused by inadequate thyroid function.[27,28] The report marked "an important advance in understanding – that a gland such as the thyroid which was known not to possess a duct was able to secrete something into the circulation, and that this secretion was necessary for health."[44]

Medical Treatment of Hypothyroidism

Once the function of the thyroid gland was understood, it was necessary to find an adequate treatment that would compensate for the missing hormone essential for life. "In 1891 George Murray (1865-1937) of Newcastle-upon-Tyne, a former pupil of Horsely, managed to prepare an extract of sheep's thyroid in glycerin, which he injected subcutaneously into Mrs. S., a 46 year-old lady with myxedema. The patient promptly improved and remained well while receiving thyroid extract for twenty-eight years."[27] In 1892 "Edward Fox (1859-1938) of Plymouth, England administered 'half sheep's thyroid, lightly fried and taken with currant jelly once a week' with marked patient improvement of the myxedematomas patients."[31] Similar success by feeding fresh thyroid glands to patients after thyroidectomy was also reported by Hector Mackenzie the same year.[32] These studies supported the concept of organo-therapy, a term coined by Brown-Sequard.[44-46] In 1893 Kocher suggested that the thyroid gland might contain iodine, and in 1896 Eugen Baumann (1846-96) of Freiburg Germany, found "that the compound iodothyrin was a normal constituent" (of the thyroid gland).[27]

The young Swiss medical student with an enlarged thyroid gland that manifested in an asymmetrical symptomatic goiter would have been fortunate to discover his condition at the time that knowledge of the thyroid gland's function was finally discovered. Thanks to the careful evaluation of his patients and the persistence of Theodor Kocher and his peers, the student would probably have been reassured by the marked improvement in Kocher's patient outcome after thyroidectomy. He would have therefore agreed to have a right thyroid lobectomy.

Theodor Emil Kocher's Contribution to Surgery

Kocher is considered by many to be the "father" of thyroid surgery.[20] He was the first surgeon to receive the Nobel Prize for Medicine for his clinical discovery and technical excellence "for his works on the physiology, pathology and surgery of the thyroid gland."[20] Kocher reestablished the traditions of Ambroise Paré (1510-1590) and John Hunter (1728-1793), basing surgery on anatomical and physiological principles.[40,42] Under his chairmanship Berne became a world center for surgery. Kocher's surgical text *Chirurgische Operationslehre* was printed in five editions and translated into many languages. He also published numerous articles, monographs, and dissertations.[33] In addition

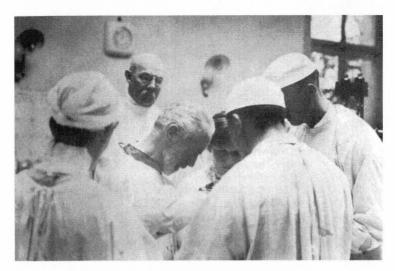

·Figure 3: Photograph of Dr. William S. Halsted (1852-1922) (Back-center) observing
Dr. Theodor E. Kocher in operating theater in Bern, Switzerland

to his contributions to thyroid surgery, Kocher also made seminal contributions regarding shoulder relocation, mobilization of the duodenum (Kocher
maneuver), and new operations for inguinal hernias and carcinoma of the
rectum. He developed new instruments such as the Kocher clamp and helped
establish the importance of basing surgery on sound pathophysiology. As
mentioned in his obituary, Kocher was recognized as being "a slow operator,
but there was never a moment wasted." "He has one speed which I designated
as the 'Kocher speed'."[47] "As a surgeon he was unsurpassed in observance of
detail in technique."[47,48]

Kocher was also recognized as a superb teacher. His skill as an artist
helped him with his teaching in the operating theater.[48] He "trained no great
men in his own school" as Billroth did, but "he helped inspire and educate
many visiting surgeons from around the world including William S. Halsted
(1852-1922) and Harvey Williams Cushing (1869-1939) from John Hopkins
Hospital, George Crile (1864-1943) from the Cleveland Clinic; Charles Mayo
(1865-1939) from the Mayo Clinic; Rene Leriche (1879-1955) from Strasbourg
and Lyon, France, C. Roux (1857-1934) from Lausanne, Switzerland and other leading surgeons.[33] William S Halsted (1852-1922) wrote that Kocher was
"one of the best technical surgeons of all time and it was his technique that
Halsted adopted."[21] Halsted stated "many times in the past 20 years I have

stood at the side of Professor Kocher at the operating table enjoying the rare experience of feeling in quite complete harmony with the methods of the operator."[21] (Figure 3) Kocher's "operations numbered many thousands including over 5,000 thyroidectomies".[21] His scrupulous and minutely careful surgical technique were unsurpassed; "every tiniest detail was arranged, every difficulty most gently overcome; there was not haste, no untidiness, no shedding of one drop of blood that could be spared. Infinite accuracy, intimate care, infinite patience, gave him results as near to absolute perfection as it is possible for surgery to go."[47,48] In short, Kocher was a "complete surgeon."[21] The field of endocrine surgery is indebted to Theodor Kocher for his myriad contributions.

References

1 Welbourn RB. The History of Endocrine Surgery. New York: Praeger Publishers; 1990: 21.
2 Vesalius, A. de Humani Corporis Fabrica, Libri Primus Bruxellensis, 1543:367.
3 Welbourn RB. The History of Endocrine Surgery. New York: Praeger Publishers; 1990: 20.
4 Merke F. History and Iconography of Endemic Goitre and Cretinism. Lancaster/Boston/The Hague/ Dordrecht: MTP Press Ltd.; 1984:152-153.
5 Merke F. History and Iconography of Endemic Goitre and Cretinism. Lancaster/Boston/The Hague/ Dordrecht: MTP Press Ltd.; 1984:234.
6 Clark C and Clark OH. Remarkables: Endocrine Abnormalities in Art. Berkeley: Univ of Calif Press; 2011:19.
7 Merke F. History and Iconography of Endemic Goitre and Cretinism. Lancaster/Boston/The Hague/ Dordrecht: MTP Press Ltd.; 1984:185.
8 Ahmed AM and Ahmed NH. History of disorders of thyroid dysfuncation. *Eastern Med Health J* 2005;11:459-469.
9 Welbourn RB. The History of Endocrine Surgery. New York: Praeger Publishers; 1990:7.
10 Garrison FH. An Introduction to the History of Medicine. Fourth Edition. Philadelphia: W.B. Saunders; 1960:505.
11 Bigelow HJ. Insensibility during surgical operations produced by inhalation. *Boston Med Surg Jour* 1846; 35(16):309-17.

12 Garrison FH. An Introduction to the History of Medicine. Fourth Edition. Philadelphia: W.B. Saunders; 1960:506.

13 Lister J. New Method of treating compound fracture, abscess, etc. *The Lancet* 1867; 90(2291):95-6.

14 Garrison FH. An Introduction to the History of Medicine. Fourth Edition. Philadelphia: W.B. Saunders; 1960:591.

15 Wells S. Reports of Medical and Surgical Practice in the Hospital of Great Britian Britain. *Br Med J* 1874;1:47

16 Welbourn RB. The History of Endocrine Surgery. New York: Praeger Publishers; 1990:8.

17 Welbourn RB. The History of Endocrine Surgery. New York: Praeger Publishers; 1990:6.

18 Welbourn RB. The History of Endocrine Surgery. New York: Praeger Publishers; 1990: 27.

19 Suskind A. Ueber die Estirpation von Strumen. Germany: Inaugural-Abhandlung, Tubingen; 1877.

20 Becker WF. Presidential address: Pioneers in thyroid surgery. *Ann Surg.* 1977; 185(5): 493–504.

21 Harwick RD. Our legacy of thyroid surgery. *Am J Surg.* 1988;156(4):230-4.

22 Felger EA, Zeiger MA. The death of an Indian Rhinoceros. *World J Surg.* 2010 Aug;34(8):1805-10.

23 Welbourn RB. The History of Endocrine Surgery. New York: Praeger Publishers; 1990:217-218.

24 Gley E. Sur les Fonctions du corps thyroide. *Comptes Soc Biol* 1891; 43:841-847.

25 Vassale G, Generali F. Fonction parathyreoldienne et Fonction thyroidienne. *Arch ital de biol* 1896;33:154-56.

26 Welbourn RB. The History of Endocrine Surgery. New York: Praeger Publishers; 1990:25.

27 Welbourn RB. The History of Endocrine Surgery. New York: Praeger Publishers; 1990:29.

28 Garrison FH. An Introduction to the History of Medicine. Fourth Edition. Philadelphia: W.B. Saunders; 1960:695.

29 Medvei, VC. History of Endocrinology. London: MTP Press, Ltd.; 1982:248.

30 Murray GR. Note on the treatment of myxoedema by hypodermic injection of an extract of the thyroid gland of a sheep. *Br Med J.* 1891;

2:796-797.

31 Fox EL. A case of *myxoedema treated* by taking *extract of thyroid* by the mouth. *Br Med J 1892; 2:941.*

32 Mackenze HWG. A case of myxoedema treated with great benefit by feeding with fresh thyroid glands. *Br Med J* 1892;2:940-941.

33 Gemsenjager E. Theodor Kocher. In Press. European Thyroid Association Arch, Milestones in European Thyroidology.

34 Welbourn RB. The History of Endocrine Surgery. New York: Praeger Publishers; 1990:35.

35 Sakorafas GH. Historical evolution of thyroid surgery: from the ancient times to the dawn of the 21st century. *World J Surg.* 2010; 34(8):1793-804.

36 Welbourn RB. The History of Endocrine Surgery. New York: Praeger Publishers; 1990:31.

37 Welbourn RB. The History of Endocrine Surgery. New York: Praeger Publishers; 1990:42.

38 Vellar IDA. Thaoms Peel Dunhill, The forgotten man of thyroid surgery. *Medical History* 1974; 18:32-50.

39 Welbourn RB. The History of Endocrine Surgery. New York: Praeger Publishers; 1990:32.

40 Kocher ET. Concerning pathological manifestations in low-grade thyroid diseases. In Nobel Lectures in Physiology or Medicine, 1901-1921. Amsterdam: Elsevier Publishing Co.; 1967:330-383.

41 Medvei, VC. History of Endocrinology. London: MTP Press, Ltd.; 1982:237.

42 Clark OH. Influence of Endocrine Surgery on General Surgery and Surgical Science. *Arch Surg.* 2009;144(9):800-805.

43 Welbourn RB. The History of Endocrine Surgery. New York: Praeger Publishers; 1990:46.

44 Welbourn RB. The History of Endocrine Surgery. New York: Praeger Publishers; 1990:33.

45 Ord WM, Horsley V, Semon F, et al. Myxoedema report. *Trans Clin Soc London* 1888;21:Suppl.

46 Ginn SR, Vilensky JA. Experimental confirmation by Sir Victor Horsley of the relationship between thyroid gland dysfunction and myxedema. *Thyroid* 2006;16(8):743-7.

47 Obituary: Professor Theodor Kocher. *Br Med J* 1917; 2:168-169.

48 Moynihan B. Obituary Professor Theodor Kocher. *Brit Med J* 1917;2:1681169.

3

The Parathyroid Glands
From Early Misperceptions to Modern Understanding

Erin A. Felger and Martha A. Zeiger

Introduction

From the first discovery of a parathyroid gland in an Indian rhinoceros by Sir Richard Owen to the development of the first parathyroid hormone extract by Adolph Hanson and J.B. Collip, numerous fascinating historical anecdotes mark the history of the parathyroid glands. This chapter traces the story of discovery, speculation, misconceptions and scientific inquiry that have led to our present-day concepts of the anatomy and function of the parathyroid glands. These stories, which have been passed on by generations of physicians around the world, are rich with intellectual pursuit and medical achievements. Sir Richard Owen in England and Ivar Sandström in Sweden were the first to identify parathyroid glands in animals and humans, respectively. In France, Eugène Gley was the first to associate parathyroid glands with tetany, but his explanation, called the "detoxification theory", was later proven to be the complete opposite of our modern understanding of parathyroid physiology. Jacob Erdheim, another brilliant physician and scientist from Austria, was the first to associate bone disease with abnormal parathyroid function. His conclusions, however, about compensatory parathyroid hyperplasia in response to bone disease were also erroneous, but went unchallenged for over a decade. Finally, in 1909, MacCallum, an American, published the first study regarding parathyroid function and concluded that the glands exerted control over calcium metabolism and that tetany in turn was due to insufficient parathyroid secretion. Further supporting the proposed pathophysiology, two American scientists, Hanson and Collip, independently developed a parathyroid extract that they showed to be useful in the treatment of tetany.

From generation to generation, these historical twists and turns have provided us with a deeper appreciation of the function of the parathyroid glands and the course taken toward our current understanding of these glands and their associated diseases.

Death of a Rhinoceros

Richard Owen began his medical studies at Edinburgh University in 1824, where he developed his love for anatomy (Figure 1). In April 1825, he accepted a position with Dr. John Abernathy, President of the Royal College of Surgeons at St. Bartholomew's Hospital in London. There he became Abernathy's prosector, preparing all of the anatomic specimens used in Abernathy's lectures. This arrangement was particularly advantageous for Owen because he obtained experience dissecting without needing to personally purchase the cadavers, a common practice for anatomists at that time. One year after his appointment, on August 18, 1826, Owen became a member of the Royal College of Surgeons, and Abernathy appointed him as assistant curator of the Hunterian Collections. This position gave Owen the unique opportunity to dissect animals that had been under the care of the Zoological Society of London [1]. Although, as the Zoo prosector, Owen is most famous for his dissection of the rhinoceros, his contributions to zoology began with the dissection of an orangutan and continued with dissections of a giraffe, an ape, a chimpanzee and several marsupials, all of which had died at the Zoo.

Figure 1: Sir Richard Owen at age 42 holding a bone of the Dinornis Maximus from *The Life of Richard Owen*. Vol.1. London: John Murray; 1894

Figure 2: A Stubbs painting of a rhinoceros from *The Rhinoceros from Durer to Stubbs 1515-1799*. Vol.1. London: Philip Wilson Publishers Ltd; 1986

When the London Zoological Society purchased a male rhinoceros (Figure 2) in 1834, they had no idea how much the death of this animal would ultimately contribute to the field of anatomy. Prior to the rhinoceros' acquisition, Owen, a prominent anatomist, had persuaded the Society that the animal would be a wise investment; little did they know how true this would become. On May 24, 1834, an Indian rhinoceros was purchased for one thousand guineas (approximately $108,000 in today's dollars) and quickly became the most popular exhibit in the zoo.[1,2] Unheard of today, but probably typical of the times, the rhinoceros was housed in the Elephant House. At some point during his stay in the Elephant House, a large male elephant in the adjacent paddock began to torment the rhinoceros by forcing him to the ground with his large tusks. Four months after the elephant began these attacks, the rhinoceros started vomiting "a bloody and frothy mucous" and died one week later on November 19, 1849.[3]

Upon his death, Owen became reacquainted with the rhinoceros he had originally persuaded the Zoological Society to obtain fifteen years prior. The rhinoceros was brought to Owen in November of 1849, and because of Owen's status within the scientific community, he was given "The very rare opportunity of investigating the internal structure of the Rhinoceros…"[4] Over the course of months, Owen performed a meticulous and comprehensive necropsy. During the dissection, Owen carefully delineated the structures within the

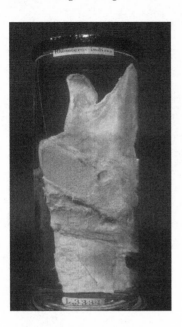

Figure 3: Owen's dissection of the parathyroid gland of the Indian Rhinoceros reproduced with permission from the Royal College of Surgeons of England

upper airway, including the thyroid. He noted, "The thyroid gland consisted of two elongate, subtriangular lobes extending from the sides of the larynx to the fourth tracheal ring… The structure of this body is more distinctly lobular than is usually seen; a small compact yellow glandular body was attached to the thyroid at the point where the veins emerge." Believing that this glandular structure was somehow significant, he preserved it *in situ* along with the thyroid and larynx. It can still be viewed today in the Hunterian Museum, located inside The Royal College of Surgeons in London. (Figure 3).[3]

A paper describing the anatomy of the rhinoceros that resulted from this meticulous dissection was presented as a lecture at the Zoological Society on February 12, 1850 (this lecture was subsequently accepted as a journal article in 1852, but the volume in which it was ultimately available was not published until 1862). Because of this delay in publication, Owen's discovery of the parathyroid glands was to remain in obscurity for several decades.[1,5] It was not until A. J. E. Cave, successor to Owen as Professor of Anatomy at the Royal College of Surgeons, discovered that Owen's paper was published in 1852, not 1862 as originally thought, that Owen's findings would be finally recognized. Prior to Cave's discovery, credit for first identifying the parathyroid glands had been given exclusively to Ivar Sandström, a medical student in Sweden. Because of this ill-fated turn of events, Owen was to receive the credit he deserved for the first documented discovery of the parathyroid glands only post-

Figure 4: Ivar Sandström, (1852-1889)

humously. Up until Cave had found Owen's original paper, all of the credit for the discovery of the parathyroid gland had been awarded to Sandström, ironically, born the same year as Owen's original paper.

Parathyroid Glands in Other Mammals

Ivar Viktor Sandström, born in 1852 in Stockholm, Sweden was the son of the Secretary of Agriculture (Figure 4). Unfortunately, he inherited a mental illness from his mother that would later precipitate in suicide in 1889 at the young age of 37.[6] Prior to his death, however, Ivar Sandström made a lasting contribution to medicine with the identification of parathyroid glands in mammals.

Sandström was a medical student in 1877 at the University of Uppsala. While working as a prosector for the Anatomy Department he identified the parathyroid gland for the first time in a dog. Captivated by this discovery, he proceeded to locate the glands in a horse, ox, dog, cat and rabbit as well as in 50 human cadavers.[7, 8] Sandström recorded the size and number of these glands and mapped their locations in each specimen. He named these glands, glandulae parathyreodeae, in his seminal paper on the topic, *Om en ny körtel hos menniskan och atskilliga däggdjur* (On a new gland in man and several mammals). In his article, Sandström stated, "About three years ago [1877] I found on the

UPSALA LÄKAREFÖRENINGS
FÖRHANDLINGAR

| Band. XV. | 1879—1880. | N:r 7 & 8. |

Fredagen den 5 Mars.

SANDSTRÖM, Glandulæ parathyreoideæ. — BLIX, Myograf. — SANDSTRÖM, Prostatapreparat. — WIDE, Lefverpreparat. — HOLM-GREN, Retinaströmmen. — Svar till *H. Cohn* om färgblindheten. — Föreningens yttrande mot *K. Wicksells* föredrag om dryckenskap.

1. Om en ny körtel hos menniskan och åtskilliga däggdjur

af

IVAR SANDSTRÖM.

För snart tre år sedan påträffade jag å sköldkörteln af en hund en liten, knappt hampfröstor bildning, som låg innesluten

Figure 5: Sandström announcement of the discovery of parathyroid glands, reproduced with permission from Anton H.M. Vermuelen and Springer Publishing

thyroid gland of a dog a small organ, hardly as big as a hemp seed, which was enclosed in the same connective tissue capsule as the thyroid, but could be distinguished there from by a lighter color."[9] He postulated that the glands were an embryonic extension of the thyroid arrested in various stages of development, and he astutely recognized that their location varied in animals and man. Sandström meticulously described every aspect of the parathyroid gland including the histology and blood supply that he carefully noted was derived from the inferior thyroid artery.[9, 10]

Sandström originally submitted his paper to a German journal that returned it with a request that he shorten the paper; Sandström felt strongly that his thirty one-page manuscript should not be truncated and so he sent it to the local medical society that subsequently published it in Transactions of the Medical Society of Uppsala in 1880 (Figure 5). In his paper, Sandström credits Remak and Virchow for seeing the glands earlier, "two authors who seemed to have traced the glands in question, although I do not want thereby to deny the possibility that even others may have observed them before me."[11,9] In 1855, Robert Remak, an embryologist from the University of Berlin, had described a parathyroid gland as being associated with the thymus and distinct from the thyroid. In 1863, Rudolf Virchow renowned pathologist at the University of Berlin, identified the parathyroid gland in one of his specimens, commenting that it was not an accessory thyroid or lymph node but a distinct structure. Neither, however, pursued their observations of the parathyroid any further.[11]

Sadly, when Sandström presented his findings at a Stockholm meeting, no one paid any attention, presumably because these structures seemed to be glands without a function and were therefore not worthy of further merit. This attitude toward the parathyroid glands persisted for the next decade.[12] Despite a promising career as a dedicated researcher and educator, Ivar Sandström was a conflicted, depressed man discouraged by the lack of recognition for his work and, by a troubled marriage. He took his own life on June 2, 1889, two years before Eugène Gley resurrected his seminal work on the parathyroid gland. Ironically, Sandström's work was lauded by many as the last discovery of an organ in humans, an honor he surely would have cherished.[6]

"Detoxification Theory and Our Misunderstanding of Parathyroid Physiology"

Ten years after Sandström's discovery, Eugène Gley exhibited a renewed interest in the parathyroid glands, mainly because of his work with experimental thyroidectomies. Born in 1857 in Vosges, France, Eugène Gley became a Professor of Physiology and held the Brown-Séquard Chairmanship at the Collège de France in Paris. His work was instrumental in defining what was necessary for rigorous research and experimentation. Specifically, he introduced the idea of extensively reviewing the medical literature before developing any new experimental design, a concept routinely practiced today.[13]

Gley noticed that dogs who had the glands of Sandström excised during a thyroidectomy developed tetany and eventually died. The term tetany was first used in 1852 by Lucien Corvisart, the personal physician of Napoleon III. With his curiosity piqued, Gley studied the literature and, specifically, Sandström's paper which helped shape his experimental parathyroid injury model in dogs.[7] His model was the first to unequivocally demonstrate that aparathyreotic dogs developed tetany and died; specifically, he proved that tetany was present only when the parathyroids were damaged or removed, even if the thyroid was left intact. He reported these findings in 1891 in an article published by the Société de Biologie.[14, 15] Concurrently in Italy, Giuilo Vassale and Francesco Generali also concluded that removal of the parathyroid glands resulted in tetany and death. Their findings certainly supported the detoxification theory which stated that parathyroid glands, by an unknown mechanism, removed toxins from the body that otherwise resulted in violent shaking or convulsions. Gley also believed the detoxification theory. He further believed that these "glandules thyroidiennes" were actually embryonic thyroid tissue, just as Ivar Sandström had, and believed that their function was related to

Figure 6: Jacob Erdheim, (1874-1934)
reproduced with permission from Ruth
Koblizek and Bildarchiv der Medizinischen
Universität Wien

the thyroid gland. He stated, "a true functional association may possibly exist between the thyroid and parathyroid glands."[10, 13]

Although his theory was incorrect, Gley was the first person to assign a vital role to the parathyroid glands, previously believed to be nonfunctional. Because of the work accomplished by Gley, Vassale and Generali, surgeons now began to respect the parathyroid glands and better protect them when performing thyroid surgery. In the new century, Jacob Erdheim would continue to perpetuate these misconceptions of parathyroid function while trying to further delineate the role parathyroid glands played in bone disease.

Erroneous Theory: Calcium Pathophysiology, Parathyroid Glands and Bone Disease

In 1906, Jacob Erdheim, (Figure 6) a Viennese pathologist, confirmed Gley's work with his own studies on the effects of total parathyroidectomy in rats. He was to advance the understanding of parathyroid physiology even further by making the first observations about the long-term effects of chronic parathyroid insufficiency, thereby establishing an association between the parathyroid

a

b

c
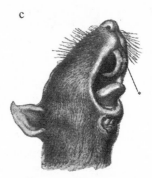

Fig. 25. Ratte mit normalen Nagezähnen.

Fig. 26. Tetanieratte. Obere Nagezähne ausgefallen, beginnen eben nachzuwachsen. Untere Nagezähne enorm lang, erzeugen ein Druckgeschwür am harten Gaumen und weisen an der vorderen Fläche in gleicher Höhe einen weißen Fleck auf.

Fig. 27. Tetanieratte. Untere Nagezähne ausgefallen. Am Unterkiefer ein Geschwür, in dem die Stümpfe der nachwachsenden unteren Nagezähne sichtbar sind. Die oberen Nagezähne abnorm lang, tragen in gleicher Höhe an der vorderen Fläche einen weißen Fleck.

Figure 7: Erdheim's rat experiment circa 1905, reproduced with permission from Anton H.M. Vermuelen and Springer Publishing

glands, bone disease, and calcium metabolism. His experiments consisted of removing the parathyroid glands in rats in order to induce tetany.[7,8] After parathyroidectomy, he noticed that the rat incisors that normally grew continually became abnormally short. Under the microscope he saw that the teeth had become decalcified. In the second phase of his ingenious experiments, parathyroid tissue from other rats was transplanted into the "parathyreoprivic" rats and again the incisors were examined. He documented that the teeth now had a layer of normal calcium deposition followed by a layer of decalcification and finally, another layer of normal calcium after transplantation of parathyroid tissue (Figure 7). Unfortunately, he erroneously concluded that the bone disease was the primary problem and not, as we now know today, secondary to parathyroid abnormalities.[11,13]

Instead, Erdheim believed that parathyroid hyperplasia was secondary to bone disease, having performed autopsies in three patients with enlarged parathyroid glands and osteomalacia. He theorized that the hyperplasia was a compensatory mechanism that served to increase calcium and reverse bone disease. During this time period, two other physicians, Max Askanazy and Friedrich Schlagenhaufer, commented on single parathyroid tumors associated with bone disease in three different cases. Shlagenhaufer, however, correctly believed it was the parathyroid tumor that caused the bone disease and he suggested that these parathyroid tumors should be surgically removed. Once again, a worthwhile recommendation was ignored for nearly a decade.[8, 13]

Because Erdheim was highly regarded within the medical field, his misconceived theory was widely accepted. Physicians further extrapolated his theory to explain osteitis fibrosa cystica as well as osteomalacia and rickets. Interestingly, Erdheim never confirmed his theory scientifically. Indeed, patients with bone disease were inappropriately treated with parathyroid extract; those who believed Erdheim's theory presumed that increased parathyroid hormone would aid in reversing the bone disease.[10, 16] The practice of providing parathyroid extract to patients with severe bone disease was to continue until two researchers back in the United States finally elucidated the mechanism behind tetany, proving it with a series of thoughtfully executed experiments.

Understanding Tetany, Finally

William George MacCallum was a member of the first graduating class of Johns Hopkins Medical School in 1897. As a pathologist he first worked at Johns Hopkins Hospital, then Columbia University for several years before returning to Johns Hopkins as Chair of the Department of Pathology. In 1903, he began performing studies on patients with goiters, examining tetany and its resolution after the administration of parathyroid extract. Initially, MacCallum believed in the detoxification theory. He stated that parathyroid glands must make "motor neurons less susceptible to circulating toxins". However, in 1909 he and Carl Voegtlin, his research partner, hypothesized that it was the calcium level that was responsible for hypoparathyroid tetany and began experimenting on parathyroidectomized dogs.[11, 17] Their experiments included the administration of various salts, including calcium, magnesium, sodium and potassium to dogs with tetany. They discovered that only calcium relieved tetany. They also astutely noted that potassium exacerbated the symptoms of tetany and that magnesium was toxic to the dogs. Their conclusions: " the parathyroids control in some way the calcium metabolism so that upon their removal a rapid excretion, possibly associated with inadequate absorption and assimilation deprives the tissues of calcium salts."[18]

To remove any doubt about the newly discovered relationship between calcium and tetany, MacCallum performed a second set of confirmatory studies. First, he dialyzed dogs to create calcium-free blood that he injected into tetanic animals. He found that the calcium-free blood had no effect on their tetany, whereas when he injected normal blood into the same animals, their tetany resolved. This beautifully designed experiment proved without a doubt that tetany was indeed caused by hypocalcaemia.[19, 20] His findings, however, were not accepted for several years, during which time other physicians and scientists continued to publish articles supporting the detoxification theory. Interestingly, MacCallum himself was also not convinced of the validity of his studies until 1924 when he definitively stated that tetany was the direct result of decreased calcium. What convinced him appeared to be the fact that William S. Halsted, working closely with MacCallum, used calcium salts successfully on his patients who developed tetany after thyroid surgery.[20, 21] Around the same time period, two men were simultaneously developing a parathyroid extract to treat tetany, going yet a step further than MacCallum in continuing the progress towards understanding parathyroid disease.

Two Scientists, One Parathyroid Extract

Adolph M. Hanson took an unusual path after receiving his medical degree at Northwestern University. He trained in neurosurgery, serving under Harvey Cushing in the U.S. Army Medical Corps during WWI. When he returned to his hometown of Faribault, Minnesota after the war he established a private practice and set up a laboratory in his basement for scientific experiments. He was a classic physician/scientist who developed experiments simply for the love of scientific inquiry (Figure 8). This path, however, was soon to clash with the newly emerging career scientists who had academic and industrial laboratories and tremendous funding at their disposal.[11, 22]

Hanson became interested in the parathyroid glands during his time with Cushing because of his familiarity with thyroid surgery and the dreaded complication of tetany. He decided to study the chemical composition of parathyroid tissue and potentially isolate its active component in hopes of finding a better treatment for tetany. In 1922, using bovine parathyroid glands, he ran several experiments. He discovered that in order to release its active component, the parathyroid gland had to be boiled in a solution of strong mineral acid.[22, 23] He developed an extract he called Hydrochloric X that had a heretofore unknown, active, organic compound in it, namely the parathyroid hormone extract. It was difficult to study the extract because Hanson could not keep laboratory animals in his home and needed help from an academic lab to carry out the next set of experiments. A.B. Bell, a researcher at the University

Figure 8: Adolph M. Hanson
(1880-1959)

of North Dakota agreed to help. Bell treated dogs that had undergone parathyroidectomy with an injection of Hydrochloric X within twenty-four hours
of their surgery. Remarkably, tetany subsided in all of the dogs within six
hours of the injection and this normal state was maintained with subsequent
smaller doses.[24, 25] Hanson's extract was active!

Unfortunately, Hanson ran into several roadblocks in his attempt to find
a company that would produce his extract. As a lone researcher working out
of his home, Hanson had neither the credentials nor the capital to catapult
him into the world of business and mass production. In November of 1924,
he contacted Eli Lilly pharmaceutical company but was told that they were
already purifying a similar extract developed by another group.[22] Who had
beaten Hanson to the finish line?

James Bertram Collip was an investigator and pioneer in biomedical research. He epitomized the career scientist with a large, well-funded and pro-
lific academic laboratory (Figure 9). He devoted his studies to endocrine function and the treatment of endocrine diseases. In 1924 at the University of
Alberta, he also had begun studying the parathyroid gland. Collip believed
that the parathyroid gland secreted a hormone that regulated calcium. Because

Figure 9: J.B. Collip, circa 1930

of his success in isolating insulin, he was able to apply this knowledge to the development of an active parathyroid hormone extract. One of his greatest strengths was the ability to extract and purify active proteins, and it was show-cased best with the development of the parathyroid hormone extract.[22, 26]

Collip began to run a series of biochemical tests on parathyroid glands that included boiling the glands in a five percent hydrochloric acid solution and then purifying the extract. He then showed through a series of experiments that tetany in parathyroidectomized dogs could be reversed using the extract. These experiments were completed with a large number of dogs as Collip had extensive resources at his disposal, unlike Hanson who worked in his basement laboratory.[22] Once he established that his extract controlled tetany, Collip took it a step further and demonstrated that serum calcium levels increased concomitantly with the reduction of tetany. Interestingly, he also demonstrated that giving his extract to normal animals increased their serum calcium level, which if elevated enough, could lead to death. During this time Collip and his research associates realized the importance of accurate serum calcium levels for their work, so they also developed a quantitative method to measure serum calcium, thereby solidifying their notoriety in perpetuity.[22, 26]

Although Collip was most interested in the science behind the extract preparation, he was also mindful of its therapeutic benefit. He successfully enlisted Eli Lilly regarding the parathyroid hormone project. Collip now not only had his well-funded research lab, but also a major pharmaceutical company supporting his research on parathyroid hormone extract.[22, 27]

Conversely, Hanson had none of these resources and was incensed when he learned that Collip had both beaten him to Eli Lilly and, was publically taking claim for the discovery of parathyroid extract. Hanson had four dogs to run his experiments; Collip had 250. Hanson was also outraged because when Collip presented his work at various conferences he barely recognized Hanson's work: "…the papers of Dr Hanson have come to the attention of the writer. Hanson's attempts to prepare an active extract of the parathyroid glands are worthy of great commendation."[22]

Collip went on to become a leading medical researcher in Canada working with different pharmaceutical companies on hormonal research. Hanson continued to work in the Hanson Research Laboratory as a self-described patron, chief chemist, assistant chemist, stenographer, bottle washer and janitor.[22] Ultimately, both Collip and Hanson contributed to the development and understanding of parathyroid hormone. Collip received all of the recognition from the scientific world that cited only his work. However, Hanson was finally awarded the greatest accolade by receiving the patent in 1932 for the parathyroid hormone product and its extraction process.

Conclusion

From Owen's rhinoceros to Collip and Hanson's parathyroid extract, our understanding of the function of parathyroid glands has grown exponentially from their initial discovery to the modern-day treatment of parathyroid disease. This journey of scientific and intellectual pursuit has not only enriched our understanding, but advanced the management and treatment of related diseases. It serves as a quintessential story involving a small endocrine organ and its discovery, followed by the fascinating and often confused elucidation of its function.

A previous version of this chapter was published in *World Journal of Surgery* (2010 Aug;34(8):1805-10). Reprinted with permission.

References

1   Felger EA, Zeiger MA. The Death of an Indian Rhinoceros. *World Journal of Surgery* 2010; 34(8):1805-1810.

2   Officer LH, Williamson, S.H. Computing "Real Value" Over Time With a Conversion Between U.K. Pounds and U.S. Dollars, 1830 to Present 2009. Available at: http://www.measuringworth.com/exchange/. Accessed August 6, 2009, 2009.

3   Owen R. On the Anatomy of the Indian Rhinoceros. *Transactions of the Zoological Society of London* 1862; 4:31-58.

4   Owen R. The Life of Richard Owen. Second ed. Vol. 2. London: John Murray, 1894.

5   Cave AJE. Richard Owen and the Discovery of the Parathyroid Glands In: Underwood EA, ed. Science, Medicine and History, Essays of the Evolution of Scientific Thought and Medical Practice, Written in Honour of Charles Singer Vol. 2. London: Oxford University Press; 1953:pp. 217-222.

6   Rolleston H. Ivar Victor Sandstrom and the Parathyroids. *The British Medical Journal* 1938:851.

7   Eknoyan GM. A History of the Parathyroid Glands. *Americna Journal of Kidney Diseases* 1995; 26(5):801-807.

8   Organ CJ. The History of Parathyroid Surgery, 1850-196: The Excelsior Surgical Society 1998 Edward D Churchill Lecture. *Journal of the American College of Surgeons* 2000; 191(3):284-299.

9   Seipel CM. On a New Gland in Man and in Several Mammals *Bulletin of the Institute of the History of Medicine* 1938; 6(4):192-222.

10 Roher HD, and Schulte, K.M. History of Thyroid and Parathyroid Surgery. In: Oertli DaU, R., ed. Surgery of the Thyroid and Parathyroid Glands, Vol. 1. Verlag, Berlin, Heidelberg: Springer:pp. 1-12.

11 Carney JA. The Glandulae Parathyroideae of Ivar Sandstrom: Contributions from Two Continents. *The American Journal of Surgical Pathology* 1996; 20(9):1123-1144.

12 Weir N, Rimmer, J., Giddings C.E.B. History of parathyroid gland surgery: an historical case series. *The Journal of Laryngology and Otology* 2009; 123:1075-1081.

13 Vermeulen AHM. The birth of endocrine pathology. *Virchows Archiv* 2010; 457(3):283-290.

14 Gley E. Glande et glandules thyroides du chien (in French). *C R Soc Biology* 1893; 45:217-219.

15 Gley E. Nouvelle preuve de l'importance fonctionelles des glandules thyroides (in French). *C R Soc Biology* 1893; 45:396-400.

16 Kauffman. Gordon L. Jr. H, Deborah A. Historical Perspective of Parathyroid Disease. *Otolaryngology Clinics of North America* 2004; 37:689-700.

17 Longcope WT. Biographical memoir of William George MacCallum 1874-1944. Biographical Memoirs, Vol. 23: National Academy of Sciences; 1914:pp. 337-364.

18 MacCallum WG. On the relation of the parathyroid to calcium metabolism and the nature of tetany. *Bulletin of the Johns Hopkins Hospital* 1908; 19:91-92.

19 Dolev E. A Gland in a Search of a Function: The Parathyroid Glands and the Explanations of Tetany 1903-1926. *The Journal of the History of Medicine and Allied Sciences* 1987; 42:186-198.

20 MacCallum WGaV, C. On the Relation of Tetany to the Parathyroid Glands and to Calcium Metabolism.118-151.

21 Halsted WS. Auto- and Isotransplantation, In Dogs, of Parathyroid Glandules. 1908:175-201.

22 Li A. J.B. Collip, A. M. Hanson and the Isolation of the Parathyroid Hormone, or Endocrines and Enterprise. *The Journal of the History of Medicine and Allied Sciences* 1992; 47:405-438.

23 Hanson AM. An Elementary Chemical Study of the Parathyroid Glands in Cattle. *The Military Surgeon* 1923:280-284.

24 Hanson AM. Parathyroid Preparations. *The Military Surgeon* 1924:554-560.

25 Hanson AM. Standardization of Parathyroid Activity. *Jounal of the American Medical Association* 1928; 90(March 19):747-748.

26 Noble RL. Memories of James Bertram Collip. *Canadian Medical Association Journal* 1965; 93:1356-1364.

27 Collip JB, And Leitch, D.B. A Case of Tetany Treated With Parathyrin. *The Canadian Medical Association Journal* 1925 59-60.

4

William Warren Greene, 1831 – 1881
America's First Thyroid Surgeon

Walter B. Goldfarb

The early history of surgery in America is writ large with the names of the most renowned professors from prestigious medical schools in Philadelphia, New York, Boston and Baltimore. Among these men, (and they were all men), a young man achieved international recognition as an innovative bold surgeon, all the while living, practicing, and teaching in Portland, Maine.[1]

William Warren Greene was born on a farm in North Waterford, Oxford County, Maine, in 1831 (Figure 1). He began his medical studies under the preceptorship of Dr. Seth Hunkins in Waterford. He then attended lectures at the Berkshire Medical School in Pittsfield, Massachusetts, and graduated from the University of Michigan Medical Department in 1855. As noted by his junior colleague and surgical successor Frederic Henry Gerrish, "He (Greene) was offered the demonstratorship of anatomy by his alma mater, but, being dependent on his professional work for his income he was obliged to decline the position, which brought no pay but honor."[2]

Figure 1: William Warren Greene, 1872, at the time of the opening of the Maine General Hospital, Portland.

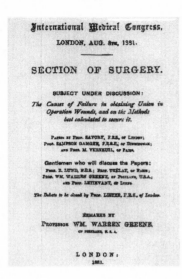

Figure 2: Program of the International Medical Congress in London, 1881. featuring William Warren Greene of Portland, Maine, as the major speaker.

Dr. Greene then moved to Waterford to practice, relocating in Gray, Maine, after 2 years. He served as a volunteer surgeon in the Federal Army for 2 months in 1862 and then returned to the Berkshire Medical College in Pittsfield after being offered the chair of Theory and Practice of Medicine.

It was noted that "he was soon transferred to the chair of surgery for which his natural endowments and cherished tastes peculiarly fitted him."[2] He accepted the professorship of surgery at the Medical School of Maine in Portland and gave his first course in 1866. He was recognized as an outstanding surgeon, one of the only 6 surgeons elected to represent this country at the Surgical Congress in Philadelphia at the Centennial in 1876. He was a featured speaker at the International Congress in London in 1881, gaining top billing on the program over others, including the renowned Sir Joseph Lister (Figure 2).

Dr. Greene was one of the founding physicians of the Maine General Hospital (today's Maine Medical Center) in Portland in 1872. In the 1870 advertisement for students for the fifty-first annual 16-week course of lectures at the Bowdoin College Medical School (the Medical School of Maine) in nearby Brunswick he is listed as professor of surgery (Figure 3). He also served as professor of surgery at the competing proprietary Portland School of Medical Instruction, whose 1871 advertisement reflects the rivalry for student enrollment (Figure 4). This was also a major source of income for the faculty. His obituary in 1881 stated that he was "a born physician and surgeon. He seemed to have a peculiar intuition as to what was wrong with the patient as soon as he looked at him." This was, of course, before the discovery of radiography and

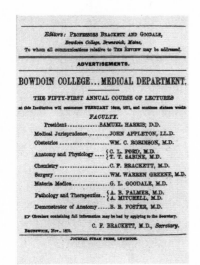

Figure 3: Advertisement in 1870 for students to enroll in the "fifty-first annual course of lectures" offered by Bowdoin College Medical School.

Figure 4: Rival Portland School of Medical Instruction's 1870 course offerings, with fees and advantages over its competitor.

the plethora of diagnostic tools that were still a few decades away. Dr. Greene contributed to the medical literature of the day on a variety of subjects. His interests and publications varied from obstetrics and gynecology to orthopedic and ophthalmic surgery, as well as reflections on medical education and practice issues. In 1866 he wrote the lead article (on successful cesarean section with survival of mother and child) in the first edition of the newly revised *New England Journal of Medicine*.[3] However, his seminal contributions to the early surgery of the thyroid gland for goiter disease are what distinguished Dr. Greene and attained for him worldwide recognition.

In 1866, Dr. Samuel D. Gross, the professor of surgery at Jefferson Medical School in Philadelphia and one of the major figures in American surgery in the nineteenth century, asked rhetorically, in his famous textbook *A System of Surgery*, the bible of surgery at the time.[4]

> Can the thyroid gland when in a state of enlargement (goiter), be removed with a reasonable hope of saving the patient? Experience emphatically answers no. No sensible man will, on slight consideration, attempt to extirpate a goitrous thyroid gland…Every stroke of his knife will be followed by a torrent of blood, and lucky will it be for him if his victim lives long enough to enable him to finish his horrid butchery…This operation is drawing of

rebuke and condemnation. No honest and sensible surgeon, it seems to me, would ever engage in it.

These were strong words uttered by the major figure in American surgery at the time. Previously, this negative attitude toward operating on the thyroid gland for goiter had been strongly noted by Liston, who in 1846 stated, "you could not cut the thyroid gland out of the living body in its sound condition without risking the death of the patient from hemorrhage. It is a proceeding by no means to be thought of".[5] This was written in the year (1846) that ether anesthesia was introduced. The French Academy of Medicine, in 1850, prohibited operations on the thyroid because of the complications of hemorrhage and sepsis. As will be seen, Dr. Greene was a most prudent, honest, and judicious surgeon. He did not enter into surgery of the thyroid gland lightly or without considerable thought.

At this time, 1866, general anesthesia had been in use for 20 years; however, antisepsis and the development of the hemostat, used for the control of bleeding, the bête noir of surgeons, were still a few years off. By today's standard, the equipment and facilities were primitive.

The story of surgery of the thyroid gland, perhaps as with no other organ, has provided a metaphor for the development of modern surgery. As William Stewart Halsted, the renowned chief of surgery at Johns Hopkins and founder of surgical training in America in 1920 stated:[6]

> The extirpation of the thyroid gland for goitre typifies, perhaps more than any other operation, the supreme triumph of the surgeon's art ... Is there any operative problem propounded so long ago (more than a thousand years) and attacked by so many which has cost so much thought and endeavor and so many lives before its ultimate solution was achieved...and further is there any problem in surgery having required for its solution such intrepid throbbing and prolonged striving of the world's greatest surgeons which has yielded results so bountiful and so adequate?... in the story of development of the operation for goitre the essential history of surgery is comprised.

Calling the presence and treatment of symptomatic goiter disease "a perpetual challenge to the physician and to the surgeon a stigma as well," Halsted provided a detailed chronologic history of the surgery for goiter in the Western world-surgeon by surgeon, country by country-recording priorities and advances. Included in this history are some of the greatest European surgeons of the time: Theodor Kocher, Theodor Billroth, Anton Wolfler, J Mikulicz, PH Watson. Among the nineteenth century American surgeons, William Warren Greene of Portland, Maine is accorded special distinction by Halsted. Greene's pioneering contributions to the surgery of goiter beginning in 1866 are cited 18 times by Halsted in his classic essay, "The Operative Story of Goitre."[6]

It is of interest that in the subsequent multiple histories of surgery of the thyroid gland in the American surgical literature since Halsted's magnum opus, Dr. Greene's name has not appeared. He is to the history of thyroid surgery in America what the Australian surgeon Thomas Peel Dunhill is referred to by his countryman Vellar as "the forgotten man of thyroid surgery."[7] In fact, Vellar's extensive review of the history of thyroid surgery in the world fails to mention Dr. Greene's early contributions. However, two relatively recent articles regarding the surgical technique of thyroidectomy credit Greene as a pioneer in this operation, albeit over 100 years after the fact.[8, 9]

Among Greene's contributions to the surgical literature between 1866 and 1870 are extensively detailed reports of 3 operations for "formidable goiters" that were life-threatening and declared hopeless by other physicians and surgeons.[10, 11] All were successfully accomplished and the patients cured. Reporting the successful excision of these large goiters, Greene stated "I have also the satisfaction ...and cannot but believe that a careful examination of these cases will convince my professional brethren that excision of an enlarged thyroid gland is, under the circumstances which I describe, a legitimate operation;" thus establishing the propriety of the operation in what he felt were selected situations. In fact, writing in 1866[10], he stated "I have quite carefully examined the literature of this subject, and so far as I can learn, all bronchoceles that were ever successfully removed (and there were very few) were small. I am very confident that none were as large as that which I shall describe." (Figure 5).

He acknowledged the dire warnings of Samuel Gross stating "the language is certainly unequivocal" and expressed his profound respect and "high esteem" for the famous professor and his reluctance to differ with him. However, Greene took issue with Gross' assertion that "no honest and sensible surgeon ...would ever engage in it." He further noted "with the surgeon, as with the physician, his first duty is always to his patient" and in extreme cases with the patient "understanding all the facts, it is the duty of the surgeon to give him that chance." In the introductory paragraph of his first case report Greene acknowledged the risks and dangers involved in undertaking such a major surgical procedure.[10]

Figure 5: Woodcut of Dr. Greene's first patient with a "formidable goiter"—preoperative and postoperative-to illustrate a successful outcome. The weight of the tumor was 1 pound, 9 ounces. "The entire operation occupied twenty-two minutes".[10]

It is well understood by the members of the profession that extirpation of an enlarged thyroid gland is one of the most fearful operations ever undertaken by the surgeon. While there is always great danger from shock, secondary hemorrhage, inflammation of the cervical vessels and of the esophagus and respiratory organs, *the* danger which overshadows all others, hanging like a thunderbolt over patient and operator, is terrible and uncontrollable hemorrhage; and while each of the other causes has in its turn produced fatal results, the great majority of deaths have occurred from that last named. In fact, so extremely vascular are these growths, both from the great enlargement of the original vessels, and also, in some cases, from an almost infinite multiplication of their branches and anastomoses, that even in small tumors, no larger than a hen's egg, such surgeons as the Coopers, the Bells, Ferguson, Velpeau, and others, have been obliged to abandon the operation; and in more than one instance the patient has died upon the table before the growth could be removed or the bleeding arrested.

In his discussion, he recounts his apprehension and details the reasons for surgery. As a wise and thoughtful, but bold, surgeon submitting his work for peer review, he indicated that he considered surgery for goiter only as a life-saving measure, as a last resort:[11]

I prefer to submit these cases to the profession with very little comment. They are the only ones in which I have ever attempted the excision of the bronchocele (goiter) and if they are the last I shall not regret it for while their issue has been so fortunate, I am sure that no man could witness even much less perform these operations, and envy the man whose lot it fell to undertake them. Yet under similar circumstances I should not shrink from such responsibility, this for the reason that the possibility of successful extirpation, even of the worst cases is established; and I believe the operation performed in the manner I have indicated, may claim quite as secure a place among legitimate dernier resorts as amputation at the hip joint.

The concept of informed consent, an issue pertinent in our day, was uppermost in Dr. Greene's mind when he wrote in 1870 about a patient with a large goiter:[11]

I told her in all probability the removal of the growth would be found impracticable; and that even if she survived the operation, which was not likely, the chances were a hundred to one that she would die of secondary affections (infection) but still there was a bare possibility of success, and while I would by no means advise an operation, yet if she, being fully aware of all the facts, insisted upon taking this desperate chance, I would undertake removal of the tumor. I was happy to have my own views as to the propriety of this advice endorsed by Dr. Storer. She immediately decided upon its being done, and at once.

Never flinching from the onerous task at hand and the possible adverse consequences, Dr. Greene displayed not only a compassion for the patient but also the courage and resolution that characterize the mature surgeon. His prudence and caution in breaking new surgical ground are noted as he addressed his colleagues in words that are as applicable today to all surgery as they were when written:[11]

In conclusion, I cannot refrain from one word of warning to my younger brethren whose ambition may make their fingers tingle, lest they should in the light of these successful cases be too easily tempted to interfere with these growths. It is and always will be exceedingly rare that any such interference is warrantable; never, for relief of deformity or discomfort merely, only, to save life. And if it is beyond all question, determined in any given case that such an operation gives the only chance for snatching a fellow being from an untimely grave, be it remembered that accurate anatomical knowledge, and perfect self control under the most trying ordeals through which a surgeon can pass, are indispensable to its best performance.

It was this bold pioneering surgical effort, well thought out, skillfully executed, and honestly presented to the medical profession for its reflection and consideration, that makes Dr. Greene's name stand at the forefront of American surgeons and which Dr. Halsted praised.

Writing 54 years after Greene's article and despite his unqualified praise, Halsted took issue with the specific operative technique as described by Greene and noted "the several steps of the operation, as I perform it" including[6]

...reflection and enucleation of the tumor with the fingers and handle of the scalpel, paying no attention to hemorrhage, however profuse, but going as rapidly as possible to the base of the gland and compressing the thyroid arteries. ...transfixion of the pedicle from below upward with a blunt curved needle armed with a double ligature...dividing the pedicle and ligating the arterial trunks ...and excision of the gland.

Halsted commented that Greene's surgical method "was not commendable" and noted that Wolfler, in 1879, contrasted the work of his countrymen with that of the surgeons of other lands "condemning particularly and quite properly the operative method of Warren Greene...which he termed, not inaptly, enucleation a tout prix".[6]

Despite these misgivings, multiple references to Dr. Greene's seminal contributions were noted in the 1870s and 1880s by a variety of European surgeons. Halsted goes on to attribute primacy and "honors of the early triumphs in the surgery of the thyroid gland equally between Germany, England, France, and America," mentioning the name of one surgeon for each country: Kocher of Switzerland, Watson of Edinburgh, Michel of France, and Greene of American (fairly esteemed company for a young man from North Waterford, Maine). Halsted concluded by stating that "it is hard to realize that today (1920) what it meant to do an operation of such magnitude and to recall the sensation made in this country by the relatively crude operations of the courageous Warren Greene[6]; high praise from the premier surgeon in America at the time. Whether Greene performed any further thyroidectomies in his career is not recorded.

In July 1881, Dr. Greene was the invited featured speaker at the International Medical Congress in London. The program highlighted his name, as noted previously, over the famous surgeons of the British Isles, including that of the great Lister. It is of interest that his topic was related to the failure of operative wound healing rather than surgery of the thyroid, his major claim to international surgical fame.

In a letter to his brother, George Frye Greene, dated August 17, 1881, he stated:

> The congress was all that could be expected. All the best men in Europe were here and a fair representation from America. I hope I represented our state creditably. I am very busy in hospital work now. Everybody is so kind to me. You will be glad to know that my name was familiar to all the surgeons of Europe and all have honored me in everyway.

He was invited to lecture and operate at the leading hospitals in London and Edinburgh before his ill-fated departure.

Returning home on the steamer Parthia, "he was suddenly attacked with complete suppression of urine and died in uremic convulsion, after an illness of hardly twenty-four hours".[2] There were no facilities for body preservation aboard; thus, burial was at sea on September 10, 1881, the day he died. A fellow passenger, J. L. Little, MD, from New York City, in a letter to George Greene in November 1881, stated that he had been delighted to have met Warren Greene as a shipmate both going to and returning from the conference: "I had known him by reputation for many years, but had never met him before," further attesting to his renown.

Thus ended the short spectacular surgical career of this highly esteemed practitioner, teacher, and medical thinker. Only 50 years old at his death, who knows what further contributions he might have made. This was during the last quarter of the nineteenth century, a watershed period in the history of medicine and surgery with major advances in anesthesia, hemostasis, surgical techniques, instrumentation, and, most especially, the emerging discipline of bacteriology; not to mention the improvements in the quality of medical education and practice. There is no doubt that he would have been at the forefront of these advances, as indicated in his 1881 presidential address to the Maine Medical Association, an enlightened and prescient treatise. Aware of the rapid scientific and medical progress of the last quarter of the 19th century he declared that the "great defect in the medical educational system is the Private Preceptorship."[12] He called for its abolition in favor of a more standardized rigorous scientific curriculum, including knowledge of French and German, which would result in better-educated medical practitioners than those currently produced by this antiquated system.

Dr. Gerrish, in his 1883 tribute to the memory of Dr. Greene, stated that for years his work:[2]

> both as a practitioner and a teacher, was almost exclusively in the department of surgery and in this branch of medicine he had the leading practice in Eastern New England. He was unquestionably a genius. Few men possessed his diagnostic skill, fewer his marvelous facility with the knife; indeed as an operator, it is difficult to conceive of his superior. His grace, steadiness, and rapidity so striking as to make the most difficult operation appear simple and easy. His daring was equal to his dexterity…and, while yet a young man in the profession, he performed the bold operation which, more than anything else, will hand his name down to posterity. This was the successful removal of a large bronchocele, in 1866, from a patient whose case had been pro-nounced as hopeless by a number of able surgeons.

Gerrish goes on to recount Dr. Greene's early pioneering forays into the surgery of goiter, which by this time, 1883, were still a small part of the surgical armamentarium not only in Europe but also in the United States where the complications of sepsis, hemorrhage, and damage to the recurrent laryngeal nerves had been largely replaced by those of postoperative tetany and ca-chexia struma priva.[6] Until the causes of all these problems were identified and addressed, progress in thyroid surgery proceeded at a slow pace for the next few years. As Halsted stated of his time working and observing in Vienna in 1879 and 1880.[6]

> I do not recall, however, having seen an operation for goiter in the clinic of Billroth*, which I attended quite regularly. From 1880 to 1886, the period of my surgical activities in New York, I neither saw nor heard of an operation for goiter, except that in one instance I assisted Dr. Henry B. Sands to extir-pate a small tumor from the right lobe of the thyroid gland.

---

*Billroth had earlier abandoned the procedure after his initial thyroidectomies, "18 of which were merely enucleations of circumscribed growths", resulted in the deaths of 8 of his first 20 operations. Most were the result of postoperative infections. He later "ventured to operate again upon goiters; and the acquired confidence was based chiefly on the results obtained with the antiseptic method of Lister, a gift from England."[6]

He described the procedure with the patient sitting upright and "a rubber bag to catch the blood tied about his neck." The lack of hemostatic instruments, which was a major inhibiting factor in the development of thyroid surgery in this era, is noted when he stated "we had only two artery forceps, all that the hospital afforded, and these were of the mouse-tooth or bulldog variety." These observations make Greene's achievements all the more remarkable. It was Greene, more than any other American surgeon, who established the propriety of this operation, and his early courageous surgical efforts were subsequently affirmed by the former naysayers.

It is of note that with better instrumentation and anesthesia, the rise of Listerism, as well as further understanding of the anatomy and physiology of the thyroid gland (and parathyroids as well), in the four decades subsequent to Greene's first report in 1866, Theodor Kocher won the Nobel Prize in 1909 for his understanding of and safe surgery of the gland. It was about 70 years from Gross' admonition of the dire consequences of engaging in thyroid surgery, also in 1866, to the mid 1930's when George Crile, Sr., was described graphically in his namesake son's provocatively entitled autobiography—"Sex, Surgery, Treasure, and Travel: The Way It Was"[13] –noting that "my father was of course delighted to have me back (at the Cleveland Clinic) and by now he really needed me. His eyesight had deteriorated to the point that he could not read, drive, or even walk without stumbling over things. Yet he was still doing five or six operations a day, mainly thyroidectomies and adrenal denervations, and these he was doing almost entirely by sense of touch...in retrospect, the conviction that made my father continue to operate could be criticized because it certainly caused accidents that threatened the lives of some patients."[13] Several of these "accidents" which were indeed fatal are discussed in detail. There is a photograph in the book of Dr. Crile, Sr. in the operating room doing his 25th thousand—that's 25,000—thyroidectomy surrounded by 13 people of whom seven are in street clothes and unmasked.

This evolution of thyroid surgery in America...from the warnings of its most eminent 19th century surgeon (Gross) and Greene's early cases to the scene 70 years later of another of this country's most eminent 20th century surgeons (albeit in decline) doing 5-6 operations a day while legally blind, is ironic, and may give new meaning to the term "surgical progress". There is no doubt that William Warren Greene of Maine was a pivotal figure in this saga.

While Greene may not have been the first to operate upon the thyroid gland in this country, he could well be regarded as its first thyroid surgeon and indeed, possibly America's first endocrine surgeon. Several isolated case reports of operations recorded earlier than his were for much smaller goiters and with considerable mortality rates. His recordings are also more thorough and detailed. He certainly has priority in the successful removal of large goiters. The subsequent recording of these cases in the medical literature for his peers to review reflect his candor, humility, and honesty. His ruminations and the baring of his concerns, especially pre-operatively, of all aspects of these dangerous surgical procedures surely places him in the forefront of this country's surgical pioneers, and the endorsement proposed by Halsted who noted in his historic opus on the thyroid.[6]

> The thyroidectomies of Warren Greene deserve conspicuous mention in the history of American surgery, for this reason I have quoted at such length from the picturesque and spirited describings of this dauntless practitioner...He is quite universally believed to be the pioneer in this field in America...He was a courageous and probably dextrous operator.

Gerrish concluded his tribute to his late mentor saying "the various communities in which he exercised his art with such success will long mourn him...The medical profession of his native state with gratitude acknowledges that to him more than any other in its ranks, it owes progress and knowledge, incentive to activity, and increase of achievement".[2] A fitting epitaph to an innovative bold surgeon, an inspiring prescient medical educator, and a beloved and skilled practitioner.

A previous version of this chapter was published in *Surgery* (2003 Mar;133(3):331-5). Reprinted with permission.

References

1   Goldfarb WB. William Warren Greene 1831-1881: pioneering Maine surgeon. *Surgery* 2003; 133(3):331-5.
2   Gerrish FH. Biographical sketch of William Warren Greene of Portland. *Trans Maine Med. Assoc.* 1883; 1:1-2.
3   Greene WW. Case of caesarean section: mother and child both saved. *Boston Med Surg J* 1868; 1:1-2.
4   Gross SD, Banov L. A system of surgery : pathological, diagnostic, therapeutic and operative. 4th ed. Philadelphia: Henry C. Lea, 1866.
5   Liston R, Mütter TD. Lectures on the operations of surgery : and on

diseases and accidents requiring operations. Philadelphia: Lea and Blanchard, 1846.

6   Halsted WS. The operative story of goiter: the author's operation. Johns Hopkins Hospital Report 1920.

7   Vellar ID. Thomas Peel Dunhill, the forgotten man of thyroid surgery. *Med Hist* 1974; 18(1):22-50.

8   Thompson NW, Olsen WR, Hoffman GL. The continuing development of the technique of thyroidectomy. *Surgery* 1973; 73(6):913-27.

9   Bliss RD, Gauger PG, Delbridge LW. Surgeon's approach to the thyroid gland: surgical anatomy and the importance of technique. *World J Surg* 2000; 24(8):891-7.

10   Greene WW. Successful removal of a large bronchocele. *Med Rec.* 1866; 1:441-443.

11   Greene WW. Three cases of bronchocele successfully removed. *AM J Med Sci* 1871; 11:80-7.

12   Greene WW. Private preceptorship in the study of medicine. *Boston Med Surg J* 1881; 105:25-29.

13   Crile G. The way it was : sex, surgery, treasure, and travel, 1907-1987. Kent, Ohio: Kent State University Press, 1992.

5

William Stewart Halsted and Goiter
From "horrid butchery" to "supreme triumph"

Kathleen C. Lee and Martha A. Zeiger

The Evolution of Thyroid Surgery

In 1866, America's leading surgeon of the time, Samuel D. Gross, Professor of Surgery at Jefferson Medical College in Philadelphia, cautioned his contemporaries on the perilous nature of the operation for goiter:

> No sensible man will on slight considerations, attempt to extirpate a goitrous thyroid gland. Every step he takes will be envisioned with difficulty, every stroke of his knife will be followed by a torrent of blood and lucky will it be for him if his victim lives long enough to enable him to finish his horrid butchery.[1]

William S. Halsted would write less than 60 years later in his renowned treatise, *The Operative Story of Goitre*:

> The extirpation of the thyroid gland for goitre typifies, perhaps better than any operation, the supreme triumph of the surgeon's art. ... Is there any operative problem propounded so long ago and attacked by so many which has cost so much thought and endeavor and so many lives before its ultimate solution was achieved?[1]

The juxtaposition of these two quotes highlights the tremendous advances in thyroid surgery during Dr. Halsted's prominent career that included his meticulous review of the literature and his scientific inquiries into thyroid and parathyroid physiology.

The Beginning

The story of Halsted and goiter begins shortly after he left New York Hospital, where he had served as a house physician, setting off for Europe to further his medical studies (Figure 1). In 1870, he spent a year abroad at the University of Vienna studying pathology, anatomy and embryology. There he befriended Theodor Billroth's first assistant, Anton Wölfler, and soon found himself in regular attendance at Billroth's operative clinics. Wölfler would later become the first to describe tetany after thyroid surgery in 1879, and also highlighted the danger posed to the recurrent laryngeal nerves. One morning, Wölfler, who at the time was writing his Die Entwickelung und Bau des Kropfes (The development and construction of the goiter), paid Halsted a visit in the laboratory and asked to examine Halsted's salmon sections. His interest and enthusiasm regarding the development and structure of fish thyroid glands greatly influenced Halsted and planted the seed for his remarkable career studying the thyroid gland.

Halsted and the Two Theodors

During Halsted's time abroad, Theodor Billroth in Austria and Theodor Kocher in Switzerland were at the forefront of thyroid surgery not only in Europe, but also in the world. Although in the previous decade Theodor Billroth had an operative mortality of 40% for thyroidectomy, his outcomes markedly improved with the advent of antisepsis; he performed 48 thyroidectomies between 1877 and 1881 with a mortality rate of only 8.3%. During the same time period, by 1883, ten years into his tenure at the University of Bern, Theodor Kocher had also performed 101 "extirpations" with an operative mortality of only 12.8%. In comparing the two surgeons, Halsted, decades later would astutely note their distinctive techniques, correlating them to their equally distinctive post-operative complications:

> Notwithstanding much speculation on the subject by various authors, it has
> not been made clear why Kocher's cases of cachexia strumipriva should have
> been so free from tetany, nor why Billroth's total extirpations should have
> been so frequently followed by tetany and should have so seldom mani-
> fested symptoms of thyroid deprivation. I have pondered this question for
> many years and conclude that the explanation probably lies in the operative
> methods of the two illustrious surgeons. Kocher, neat and precise, operat-

Figure 1: Group portrait of William Halsted and Bellevue Hospital interns, 1877. Halsted is standing 4th from the right. *Item 105272 from the Alan Mason Chesney Medical Archives.*

ing in a relatively bloodless manner, scrupulously removed the entire thyroid gland, doing little damage outside of its capsule. Billroth, operating more rapidly and, as I recall his manner (1879 and 1880), with less regard for the tissues and less concern for hemorrhage, might easily have removed the parathyroids or at least have interfered with their blood supply, and have left fragments of the thyroid. Surprising, however, is the fact that the function of the parathyroids was so seldom interfered with by Kocher, notwithstanding his careful procedure; for these little bodies were entirely disregarded by surgeons until years after the discoveries of Gley (1891) and, Vassale and Generali (1896).[1]

Halsted held Kocher in the highest regard and later wrote:

Many times during the past 20 years, I have stood by the side of Professor Kocher at the operating table enjoying the rare experience of feeling in quite complete harmony with the methods of the operator, and it is a pleasure to give expression to the sense of great obligation which I feel to this gifted master of his art and science.[2]

Kocher's influence on Halsted's operative practice was clear. (Figure 2) As Dr. George Heuer, Dr. Halsted's first assistant, 13th surgical resident, and biographer described, Halsted's precision in the laboratory and in the operating room:

> The sterilization of the skin of the dog, the draping of the areas concerned, the careful handling of tissues, the control of hemorrhage, the use of the best suture material of silk, the avoidance of injury to tissues and organs, the approximation of tissues without tension- all were carried out as rigidly as in the operating rooms of the hospital. The desire to make every experiment count, to have no fatalities due to faulty technic [sic] or performance, was evident in everything he did.[3]

Antisepsis and Control of Hemorrhage

Halsted blamed the lack of advancement in thyroid surgery in the United States and parts of Europe to a universal reluctance to embrace the principles of antisepsis and hemostasis:

> [Wölfler] wonders why so few operations upon the thyroid gland had been undertaken in America and France; … He finds the answer in the better control of hemorrhage, in the antiseptic technique … The intimation that the surgeons of England, France and America were not so advanced in the art of controlling hemorrhage and in the science of treatment of wounds was fully justified. Whereas most of the better surgeons of Germany and Austria and Switzerland promptly and eagerly accepted the [antiseptic] teachings of Lister there were few in England, France or America who did so until nearly a quarter of a century later. Indeed, our surgeons were novices as compared with the Germans in the art as well as the science of surgery in those days.[1]

Halsted would later re-emphasize repeatedly the importance of achieving hemostasis:

> The value of artery clamps is not likely to be overestimated. They determine methods and effect results impossible without them. They tranquilize the operator. In a wound that is perfectly dry, and in tissues never permitted to become even stained by blood, the operator unperturbed may work for hours without fatigue. The confidence gradually acquired from masterfulness in controlling hemorrhage gives the surgeon the calm which is so es-

Figure 2: William Halsted Operating in Theodor Kocher's Clinic, Berne, Switzerland, 1911. Halsted, standing, on patient's left; Kocher, sitting, on patient's right. *Item 101965 from the Alan Mason Chesney Medical Archives.*

sential for clear thinking and orderly procedure at the operating table.[2]

While practicing in New York, Halsted designed a set of artery forceps not unlike those employed today in the operating room for the purposes of hemostasis. Samuel J. Crowe, another former resident and biographer of Professor Halsted described:

> the tip of the Halsted clamp of the New York period was snub-nosed, but the length and spread of the handles were the same as at present ... In the winter of 1888, Doctor Halsted designed the instruments he wished to use for intestinal, vascular and thyroid surgery, and had them made in Paris.[1]

At that time, however, instrument handles in England and France were made of ivory, bone, wood or hard rubber. Metal instrument handles were only beginning to emerge in Germany. As such, Halsted, always seeking innovation, had his metal instrument handles made to order.

Halsted returned to New York City from Europe in 1880. By 1876, America's leading thyroid surgeon, William W. Greene of Maine, had performed a

Figure 3: William Halsted operating in surgical amphitheater wearing rubber gloves, circa 1903. *Item 159122 from the Alan Mason Chesney Medical Archives.*

total of three thyroidectomies. However, neither antisepsis nor hemostasis had yet been embraced in the United States. Indeed, Halsted catalogued only 45 operations on goiter in the United States by 1883. Shortly after his return, Halsted was appointed to a faculty position at the College of Physicians and Surgeons in New York, where he was reunited with his former mentor, Henry B. Sands. Halsted recalled:

> From 1880 to 1886, the period of my surgical activities in New York, I neither saw nor heard of an operation for goitre, except that in one instance I assisted Dr. Henry B. Sands to extirpate a small tumor from the right lobe of the thyroid gland. The patient, a male, was operated upon in the sitting posture, with a rubber bag to catch the blood tied about his neck. We had only two artery forceps, all, probably, that the hospital afforded, and these were of the mouse-tooth or bulldog variety.[1]

Halsted eagerly adopted the antiseptic principles of Lister during his time abroad and became one of the first American proponents of antisepsis. Upon his return to the United States, he introduced mercuric chloride and carbolic acid washings of hands, surgical instruments and surgical sites in the operating room. It was Halsted who commissioned Goodyear Rubber Company to fashion two pairs of rubber gloves for Miss Caroline Hampton, the head surgical nurse, whom Halsted would later marry. These were made to protect her hands from the harshness of the disinfectants. In the 1890s, the use of rubber gloves became standard surgical practice at the Johns Hopkins Hospital because they protected not only medical personnel from harsh antiseptic solutions, but also patients from the microorganisms found on the hands of surgeons and nurses (Figure 3). Halsted dedicated his career to the perfection of aseptic surgical technique which, as described by William G. MacCallum, Halsted's personal friend and biographer, "was rigorous and even painful to the staff if not to the patient." [4]

Dr. Welch's Invitation

Johns Hopkins University opened in 1876 and its first president, Dr. Daniel Coit Gilman, modeled it after Europe's renowned graduate universities. It was the first university in the United States to focus on graduate education and research. Dr. John Shaw Billings planned and oversaw the construction of the Johns Hopkins Hospital, which opened in 1889. Billings and Dr. William H. Welch, both avid proponents of medical education, then took seriously the responsibility of selecting the hospital's inaugural faculty members. Welch had been recently appointed the first Chief of Pathology of the hospital in 1884. Welch would later be appointed the first Dean of the Johns Hopkins School of Medicine, which opened in 1893. In 1886, Welch invited Dr. Halsted to join him in Baltimore to work in his research laboratory during the construction of the Hospital (Figure 4). In 1890, Halsted was named surgeon-in-chief.

Replicating Munk's Experiment

Soon after his arrival, Halsted set forth upon a series of animal experiments that included dogs, cats and rabbits. "In the autumn of 1887, at the suggestion of Dr. Welch, I began experiments on extirpation and transplantation of the thyroid gland in dogs ...."[1] Halsted had been intrigued by the findings of Professor Hermann Munk, a Prussian physiologist appointed at the Veterinary

Figure 4: Johns Hopkins Pathology Building Laboratory circa 1889. *Item from the Alan Mason Chesney Medical Archives.*

College in Berlin.

> In 1887 Munk made the remarkable statement that dogs survived experiments which deprived them of the function of the thyroid gland ... Of the several unsatisfactory interpretations of the results of Munk's experiments which suggested themselves to me was one which promised for these experiments an important practical bearing upon surgery, and induced me soon after the appearance of Munk's paper to repeat them.[1]

Thus, Halsted soon found himself driven to uncover,

> what part of both thyroid lobes a dog requires, we have not [yet] tried to determine. It is probably never precisely the same, and may depend somewhat upon the time allowed to intervene between the operations. One of our dogs enjoyed good health with one-eighth of the two lobes, but he died twenty-one days after the fourth operation, at which time his thyroid was reduced

to one-sixteenth of the two lobes. The fourth operation was performed six months after the first operation. For two weeks before his death, this dog had all the symptoms characteristic of total excision, including tetany.[2]

Dr. Halsted also documented the daily activities of his animal subjects. He encountered a variety of unexpected events, including dogfights and escapes, as well as the effects of hypothyroidism in pregnancy, and catalogued all of these observations in painstaking detail:

Dog 59. *The dog that lost the fight*
One dog (No. 59) remained perfectly well for sixty days after the isolation of both lobes and until killed by other dogs. His neck was so mangled by the teeth of the dogs that it was impossible to find the accessory thyroid glands or any trace of the isolated lobes. I regard this as one of the very few but undoubted instances of survival of dogs after extirpation of both thyroid lobes.[5]

Dog 104. *The escaped hypothyroid dog*
April 30th – The dog has tongue tremors and makes the licking movements with the tongue, which we have frequently observed in dogs whose thyroids have been extirpated. ...
June 24th – Hair is falling out, particularly over the eyes.
June 26th – The dog has escaped.
June 27th – Returned of his own accord. The hair is very thin all over his body. He scratches himself constantly. The oedema [sic] caused by the scratching gives to the skin about the face and head somewhat the appearance of myxoedema. [5]

Dog 97. *The pregnant hypothyroid dog and her pups*
April 13th – Bitch has whelped five pups. Two alive and three dead. ... The pups which were born alive died this evening. The thyroid glands of these pups are many times (at least 20 times) larger than normal.[5]

Halsted astutely noted that another dog with intact thyroid function whelped healthy pups.

Dr. Halsted and the Operation on Goiter

For reasons unknown, Dr. Halsted's interest in the thyroid gland subsided in the years following his extensive experiments on animals. The Johns Hopkins Hospital opened in 1889, and over the next 10 years only 7 thyroidectomies were performed, 6 by Halsted and one by Harvey Cushing. According to MacCallum:

> It was about 1903 or 1904 when Dr. Halsted began to interest himself especially in the surgery of the thyroid gland, reviving in these operations the old fascination it had had for him in the days of his experiments, before the opening of the hospital. Since Baltimore is not in an area where goiter is endemic, it was particularly the treatment of exophthalmic goiter that occupied him. This, of course, occurs everywhere and patients were soon sent to him from all parts of the country for operation. We studied the tissues removed at these operations and realized the nature of the process as something very specific, and different from that found in ordinary goiters. Those produce an awkward growth of the thyroid into great masses which can obstruct the trachea and interfere with breathing, but they do not cause the distressing nervous disturbances that come with exophthalmic type. German surgeons, particularly Billroth, Kocher and Mikulicz, had been removing goiters for a long time with varying success and with various ingenious devices for avoiding serious or fatal hemorrhage.[6]

By 1907, Halsted had performed 90 operations for Graves' disease with a mortality rate of just over 2%. No surgeon had a better record in such a sizable series. By 1914, he reported in his lecture to the Harvey Society of New York that he had operated on 500 cases of Graves' disease.

Halsted and the parathyroids

With his renewed interest in surgery of the thyroid gland, in 1906 Dr. Halsted began his work on the auto- and iso-transplantation of the parathyroids in dogs. In his own words:

> It was undertaken with a view to determine the course to be pursued by the surgeon when a parathyroid gland has been accidentally removed or de-

prived of its blood supply, and in the hope that it might be attended with such success as to justify the attempt to transplant this glandule from man to man.[7]

Halsted observed one dog for fifteen months after parathyroidectomy and subsequent auto-transplantation of a 0.25 millimeter right inferior parathyroid graft into the right rectus abdominus muscle. The dog maintained good health until its autograft was removed. The dog died three months later with symptoms of hypoparathyroidism.[8] Halsted's experiments led to his March 1912 *Journal of Experimental Medicine* publication with its remarkably lengthy title: "Report of a dog maintained in good health by a parathyroid autograft approximately one fourth of a millimeter in diameter and comments on the development of the operation for Graves' disease as influenced by the results of experiments on animals." Halsted emphasized the importance of preserving the parathyroids during thyroidectomy:

> The time has arrived when the surgeon must learn to recognize at operation not only the parathyroids, but the precise nature of their blood supply. It is in the control of hemorrhage that we sacrifice the parathyroid glands. The thyroid vessels must somewhere be divided in the operation for the removal of the thyroid gland. May they be so divided and secured as not to cut off the blood supply of the parathyroid glandules. The reply to this question, of course, is impossible without definite knowledge of the blood supply of these little bodies.[2]

Halsted realized that in order to preserve the blood supply he would first need to define the anatomy of the parathyroid vasculature. To accomplish this Halsted entrusted the task to a medical student by the name of Mr. Herbert M. Evans.

Once again his attention to detail and the resulting detailed understanding of anatomy ultimately afforded him the ability to perfect his operative technique in the preservation of the parathyroid vascular supply:

> As you will see by my remarks made at the last meeting of the American Surgical Association and published in the *Annals of Surgery* for August, I tie the inferior in preference to the superior arteries and - large and small supplying the lobe are clamped but not cut, and in this way bleeding is avoided. The clamps are plunged a little way into the thyroid tissue and made to seize

the vessels at some distance from the beginning of their course on the gland. Thus the circulation of the parathyroids and the recurrent laryngeal nerves are not endangered.[2]

Halsted read his "IV. (I) Excision of Both Lobes of the Thyroid Gland for the Cure of Graves's Disease. (II) The Preliminary Ligation of the Thyroid Arteries and of the Inferior in Preference to the Superior Artery" before the American Surgical Association held in Washington D.C. on May 7th, 1913.

The Operative Story of Goitre

In 1920, Halsted published *The Operative Story of Goitre*, an extensive and comprehensive catalogue of the operative cases that shaped the history and evolution of surgery on the thyroid. In his renowned treatise he compiled an extensive case series of documented surgeries on the thyroid gland from all over the globe. MacCallum recalls:

> In his Operative Story of Goitre, Dr. Halsted describes in minute detail the successive steps in his perfected operation. From the illustrations it looks simple enough, but the enormous distension of veins all over and about such a thyroid, and the extreme delicacy of their walls made the risk of great haemorrhage a terrifying thing. Indeed, as we have said, even he found this demoralizing in the beginning, but the control of haemorrhage was his forte, and it soon lost its terrors for him.[6]

Heuer (Figure 5) remarked when reflecting upon Halsted's writings:

> In his work he was as painstaking as in everything else he did. A paper when finally completed for publications represented an enormous amount of work. Not a statement was made which did not represent a search of the literature until the original article was found and studied, for his honesty compelled him not only to make accurate statements which could be verified but to give credit to others who had contributed to the subject upon which he was writing. Accuracy of statement, however was not his only concern. He labored over the form of presentation of his material, over the arrangement of paragraphs and sentences, over words, which expressed precisely his meaning. He studied the Century dictionary a great deal; and sometimes failing to find the right word, he coined one to express his meaning. This

Figure 5: William Halsted with early associates and residents during the 25th anniversary of Johns Hopkins Hospital in 1914. Left to right, standing: Roy D. McClure, Hugh H. Young, Harvey Cushing, James F. Mitchell, Richard H. Follis, Robert S. Miller, John W. Churchman, George J. Heuer. Left to right, seated: John M.T. Finney, William S. Halsted, Joseph C. Bloodgood. *Item 101939 from the Alan Mason Chesney Medical Archives.*

use of unusual words or newly coined words, although an example of his attempts at accuracy of statement, probably confused rather than clarified his meaning for his readers, especially his foreign readers. Yet no one who reads his publications, and especially his longer papers such as "The Operative Story of Goiter," ... can fail to appreciate the amount of reading, critical thought and the time in preparation which they represent.[3]

Conclusions

William Stewart Halsted's contributions to thyroid surgery were many. Not only did he produce remarkable experimental works involving thyroid and parathyroid grafts, but he also popularized numerous surgical instruments and significantly improved operative outcomes for thyroidectomy, achieving a mortality of just above 2%. His most significant contribution, however, was his standardized thyroidectomy technique.

In 1928, Professor von Eiselberg, a former pupil of Billroth and esteemed neurosurgeon of Austria wrote to W. G. MacCallum: "Doctor Halsted was regarded by the German surgeons as one of the greatest scientific surgeons, not only of the United States but of the whole world."[2]

Figure 6: "The Proffessor [sic]" circa 1910. *Item 162623 from the Alan Mason Chesney Medical Archives*

Terms and Words of Note

Cachexia strumipriva
Kocher coined the term in reference to the mental and physical deterioration that affected many of his patients post thyroidectomy. The malady was originally named after Sir William Gull who described the condition in 1873. Four years later, William M. Ord named it "myxedema." [9]

Epithelkörperchen
German for parathyroid; literal translation, epithelium corpuscles. In his handwritten notes, Halsted often referred to the parathyroid glandules as "epithelial bodies," as well as "e.p."

Extirpation
Complete excision or surgical destruction of a body part

Struma (s); strumae (pl)
Enlarged thyroid gland

Strumectomy
Surgical excision of all or part of a goiter.

References

1. Halsted WS. The Operative Story of Goitre. The Johns Hopkins Hospital Reports. Baltimore; 1926.
2. Crowe SJ. Halsted of Johns Hopkins; The Man and his Men: Charles C Thomas, 1957.
3. Heuer GW. Dr. Halsted. *Supplement to Bulletin of the Johns Hopkins Hospital* 1952; 90(2).
4. Magner LN. A History of Medicine. 2nd ed: Informa Healthcare, 2005.
5. Halsted WS. An Experimental Study of the Thyroid Gland of Dogs, with Especial Consideration of Hypertrophy of this Gland. Baltimore: The Johns Hopkins Press, 1896.
6. MacCallum WG. William Stewart Halsted Surgeon. Baltimore: The Johns Hopkins Press, 1930.
7. Halsted WS. Auto- and Isotransplantation, in Dogs, of the Parathyroid Glandules. *Journal of Experimental Medicine* 1909; 11(1):175-199.
8. Halsted WS. Report of a dog maintained in good health by a parathyroid autograft approximately one fourth of a millimeter in diameter and comments on the development of the operation for Graves' disease as influenced by the results of experiments on animals. *Journal of Experimental Medicine* 1912; 15(3):205-215.
9. Stover C. Edema and Ascites as Symptoms of Myxedema; with Report of Cases. *New York State Journal of Medicine* 1907; 7(12).
10. Halsted WS. Surgical Papers by William Stewart Halsted: 1852-1922 The Johns Hopkins Press, 1852-1922.

6

Herr Albert Jahne
Renowned Streetcar Conductor

Shelby A Holt, Jon A. van Heerden, & Bruno Niederle

Many, if not most, lectures and book chapters on disorders of the parathyroid gland include a historical prelude outlining the course of some well-known patient and his/her surgical trials and tribulations. This glimpse into endocrine surgical history inevitably recounts for the audience/reader the fascinating, and well-known, tale of the sea captain Charles Martell and his surgical odyssey at Massachusetts General Hospital in Boston, Massachusetts, by a pioneer surgeon of that era, Dr. Oliver Cope.

Somewhat ignored, or relegated to an insignificant footnote, is the equally important and fascinating story of a Viennese streetcar conductor named Albert Jahne and his surgical treatment by Dr. Felix Mandl at the University Surgical Clinic II in Vienna, Austria.

What is the reason for this historical slight? Are sea captains more prestigious than trolley car conductors? Was Boston geographically more significant than Vienna? We ponder this and have no answers for you.

The story of Herr Jahne is fascinating and has many twists and turns. The story illuminates examples of scientific discovery, surgical courage, and patient trust that continue to awe and inspire us today.

We plan to ask, and try, to answer three specific questions:

1. Was Albert Jahne indeed the first patient to undergo a parathyroidectomy?
2. Who was Felix Mandl?
3. What was the clinical course of Albert – streetcar conductor extraordinaire?

During the years following the development of anesthesia and antisepsis in the mid-19th century until the outbreak of World War I, surgery burgeoned under such prominent figures as Theodor Billroth, Theodor Kocher, Reginald Fitz, and Harvey Cushing.[1] Specifically, thyroid surgery went from an operation "no

honest and sensible surgeon would engage in" (Samuel D. Gross, Professor of Surgery, *System of Surgery*, 1866)[1] with a mortality rate of over 40% to Kocher's neat and precise technique of thyroidectomy that reduced the mortality rate to <0.5%.[1] Parathyroid surgery, however, would develop much later.

What was known at the time of the parathyroid glands? The first observation of the parathyroid glands is attributed to Sir Richard Owen, curator of the Natural History Museum in London, conservator of the Hunterian Museum of the Royal College of Surgeons, and Hunterian Professor of Comparative Anatomy who in 1849 described "a small compact yellow glandular body attached to the thyroid at the place where the veins emerge" during dissection of an Indian rhinoceros.[2] Ivar Sandström, a Swedish medical student, is credited with the discovery of the parathyroid glands in dogs and subsequently humans.[2] Despite detailed gross and microscopic descriptions of these glands in 1880, Sandström, like Owen, knew nothing of their function until Eugene Gley, a French physiologist, identified the connection between parathyroid glands and tetany in 1892.[3] The mystery of the parathyroid glands began to unfold. In 1903, the pathologist Max Askanazy discovered a parathyroid tumor in a patient with osteitis fibrosa cystica.[4] In 1906, a Viennese pathologist, J. Erdheim, elucidated the effects of parathyroid deficiency on bones by cauterizing the parathyroid glands of rats.[4] Through examination of the parathyroid glands of patients who had died of skeletal disease, Erdheim recognized hyperplasia of the glands and popularized the widely accepted but erroneous compensatory theory of parathyroid hypertrophy secondary to increased calcium metabolism in the setting of osteitis fibrosa cystica.[4]

Was Albert Jahne indeed the first patient to undergo a parathyroidectomy?

With this background we now first encounter Dr. Felix Mandl, a Viennese surgeon who on July 30, 1925 "had the courage to break new ground by seeking a parathyroid tumor and removing it from the neck of a patient" putting the birth of parathyroid surgery in the history books.[4] Mandl himself claimed to be the "first to operate upon a tumor of the parathyroid in a patient suffering from osteitis fibrosa generalisata".[5] Is this well-known and frequently quoted claim true; was Mandl really the first parathyroid surgeon?

Now entering our story is Sir John Bland-Sutton (Figure 1), a well respected, acclaimed surgeon and pathologist some 37 years Mandl's senior. Bland-Sutton established his reputation in the field of gynecologic surgery, specifically hysterectomy, and became president of the Royal College of Surgeons of England in 1923,[6] the year Felix Mandl was a chief surgical resident. Before Mandl was born in 1892, Bland-Sutton identified a parathyroid tumor during a post-mortem examination in 1886 and later described his findings in the fifth edition of his textbook, *Tumours Innocent and Malignant, Their Clinical Characters and Appropriate Treatment.*[6] In 1909, before Mandl had begun surgical training, Bland-Sutton removed a parathyroid cyst from 38 year old Fanny Scutts in Middlesex Hospital in London.[6] Lastly and most notable in the sixth edition of his textbook (1917), Bland-Sutton describes surgical removal of a parathyroid adenoma presenting acutely with an intratumoral bleed:

> a young married woman…was on a ship in the red sea, and had great difficulty in breathing; a small rounded lump was detected in her neck below the thyroid gland. It increased in size, and the dyspnoea became so urgent that one night the patient was prepared for tracheotomy and the ship's surgeon remained by the bedside with instruments. Fortunately the swelling subsided. On her return to England, I removed the rounded body, as big as a cherry, situated below the lower angle of the thyroid gland on the left side of the trachea. It had the microscopic features of a parathyroid.[6]

Figure 2: Felix Mandl

Though predating Mandl's historical operation by at least six years, Bland-Sutton's parathyroidectomy was presumably for local symptoms and not those referable to hypercalcemia or parathyroid-related bone disease. Thus Felix Mandl should go down in history as the first surgeon to perform a parathyroidectomy to specifically cure the manifestations of hyperparathyroidism.

Who was Felix Mandl?

Felix Mandl (Figure 2) began his medical training at the University of Vienna in 1910 and graduated in 1919 after a stint in the Austro-Hungarian Army during World War I.[4] Did he serve beside or perhaps under his future patient, Albert Jahne, who also served in the Great War? Mandl's surgical training followed at the Vienna Medical School under the tutelage of Julius von Hochenegg who succeeded Carl Gussenbauer and before him, Theodor Billroth, as the chair of the Department of Surgery II.[4] Julius Hochenegg described Mandl as an "untiring, industrious, and successful scientist."[4] This description gives us insight into the man who at the age of 33 years, a green new faculty member in this day and age, performed an operation for what could be considered a controversial indication at the time. This paved the way for modern-day parathyroid surgery.

As a Jew in Vienna at that time, Mandl personally and professionally must have felt the Jewish persecution and discrimination that was the order of the day. Despite this, Mandl had an inquiring mind and was intrigued by Albert and his disease process. Despite presumed scrutiny, he was driven to ask the crucial question as to whether Albert's problem was compensatory, (the prevailing understanding), or whether it was primary in causation. This was revolutionary thinking and was not well accepted by the medical community at the time. Was this professional jealousy or an innate ethnic distrust? We shall never know. Dr Mandl's perseverance and conviction are to be admired under such adverse circumstances. In 1927, two years following his historic parathyroid operation, Mandl was awarded the Venia Legendi for Surgery, only to be stripped of this prestigious award by the Nazi regime.[4] In 1932, at age 40, he was appointed head of the Surgical Department at the S Canning-Childs Hospital and Research Institute in Vienna.[4] Despite his position, he was prevented by the unfavorable political climate from returning to Vienna following a lecture tour in England in 1938.[4] Fortunately, Mandl was able to continue his practice in Jerusalem as Professor of Surgery and Head of the Surgical Department B at the Haddassah University Hospital.[4] Finally able to return to Vienna in 1947, Mandl's Venia Legendi was reinstated following submission of 233 scientific papers.[4] The career of this formidable surgical figure ended abruptly on October 15, 1957 when 65 year-old Felix Mandl died of acute heart failure after a case of the flu.[4]

Albert Jahne's clinic course

How fortunate for us that the paths of Felix Mandl and Albert Jahne intersected. Whether serendipitous or through divine intervention, their meeting and subsequent journey through the trials and tribulations of Albert's illness is fascinating to ponder.

A photograph of Albert J, the man, remains elusive. We only know him through detailed anatomic and histologic photographs from his operations and his autopsy. Born in Vienna on August 3, 1886, little is known about Albert Jahne as a boy, young man, and soldier in World War I.[4] He appears on the historical scene as a Viennese streetcar conductor ultimately incapacitated by progressive musculoskeletal symptoms over a five year period, later as a subject of Felix Mandl's "experiment with a fortunate outcome", and finally as the victim of recurrent, progressive disease ending in his death and medicine's gain.

Beginning in 1919 at age 33 years, Albert noted weakness in his legs[4], perhaps manifesting as difficulty climbing the stairs to his position in the streetcar. Imagine the swaying of the trolley car, the frequent stops, the climbing on and off and their effect on his ever-present bone pain. Navigating for an endless stream of trolley goers must have been an incredible challenge for a presumably severely fatigued and mentally foggy Albert. Not only physically tormented, he must have been terrified by what was happening to him at such a young age and fearful of losing his job. Strange as it may seem to us, he had no access to today's powerful search engines such as Google to provide an explanation for his symptoms. By 1923, at age 38 he could no longer perform his job duties, disabled by bone pain and unable to walk independently[4]. Had he made frequent visits to the doctor to seek answers or was it only when severely incapacitated that he sought medical attention? Was he able to get the available pain medications at the time, aspirin, laudanum, codeine, or morphine? We can only imagine the depths of his suffering and despair and reflect on the number of patients we see every day with "asymptomatic disease".

In 1924 Albert was admitted as a patient to the University of Vienna and diagnosed with osteitis fibrosis cystica.[4] What was known about Albert's condition at the time? Generalized cystic fibrosis had been described by Friedrich von Recklinghausen in 1891.[3] For the preceding two decades, Erdheim's theory of "compensatory hyperplasia" of the parathyroid glands in response to the skeletal changes of generalized cystic fibrosis had prevailed. This theory was the basis of Albert's initial treatment with "parathyreodin tablets" or "Collip's extract", a concentrated parathyroid preparation. Woefully unsuccessful, Albert's symptoms worsened and culminated in a spontaneous left femur fracture that ultimately healed following placement of a plaster cast.

In 1925 he was readmitted to the hospital with lower extremity paralysis, severe bone pain, cachexia, and a white precipitate in his urine.[4] At Albert's bedside appeared Felix Mandl, a young faculty surgeon with a research interest in endocrinology, in particular Recklinghausen's disease and pathologic changes of the parathyroid glands. Again fueled by Erdheim's compensatory hyperplasia theory, Mandl transplanted four fresh human parathyroid glands from an accident victim into Albert's preperitoneal space.[4] Though not directly successful, the fact that Albert's clinical status did not change after repleting a presumed parathyroid deficiency disproved the compensatory theory and led Mandl to explore Albert's neck for a parathyroid tumor. Is there a modern day equivalent to such a triumph as Mandl disproving Erdheim's some 20 years of thinking?

What were Mandl's first thoughts regarding Albert? Did Mandl have the penetrating "clinical gaze" characteristic of physicians of the time or did he see beyond the disease to the suffering young man only six years his senior reduced to a cripple confined to a hospital bed? Albert, with nothing to lose, could only place all of his trust and hope for cure in the hands of Mandl, a surgeon about whom he knew very little. Convinced that Albert's parathyroid glands were the primary and not the compensatory culprits and that he should surgically remove the diseased gland(s), Mandl planned to do an operation that he had never done, and certainly never seen. This is in strong contrast to our daily practice of being asked by patients how many parathyroidectomies we do in a year, knowing that the answer better be over one hundred. Did he obtain signed consent from Albert? Did he carefully discuss/did he know the risks of recurrent laryngeal nerve palsy, permanent hypocalcemia and surgical failure? How many prior neck operations had he done? What did he read in preparation the night before? How was the intended operation listed on the OR schedule for that day? What was his level of anxiety as he put scalpel to skin? We think high. Playing this theoretical, yet practical, mind game makes us wonder who was the true hero in the pas de deux? Who had the most courage? We would like to suggest that both patient and surgeon were equally heroic, shared a mutual admiration for each other, and that the intersection of their lives was a predestined one.

Imagine the adrenaline rush that Mandl felt upon opening Albert's neck and identifying a parathyroid tumor! Did he call in his colleagues to see the findings? In his operative note, Mandl stated:

> Under local anesthesia…the left thyroid lobe was mobilized…suddenly, in the rim between larynx and esophagus, a dark, partly grayish nodule separate from the thyroid was located in the inferior parathyroid position close to the inferior thyroid artery and between its branches adhering to the left recurrent laryngeal nerve. The tumor required sharp dissection to free it from the trachea and the nerve. During bilateral exploration, three additional structures were identified macroscopically as the three remaining normal glands….[4]

The tumor measured 2.5x1.5x1.2 cm and was grayish white on cut section.[4] Histologically the tumor had variably sized nuclei and sporadic mitotic structures.[4] The operative and pathologic description both raise the question of parathyroid carcinoma, which may in fact explain Albert's ultimate recurrent disease, failed reoperation, and death.

Equally exhilarating must have been the cries of the jubilant Albert, pain free and mobile for the first time in over five years! In one week's time he was able to leave the hospital (a far cry from our current same-day parathyroidectomies). He progressed dramatically over the ensuing two months from being bedridden to walking with a cane and a crutch. There was x-ray evidence of bone healing and increasing bone density.[4] Other than a bout of kidney stones in February 1926, Albert had a respite from his disease.

One wonders what happened with Albert during his symptom-free period. Did he go back to his job? Did he have medical follow up with Dr. Mandl or others? Whatever he did, it must have felt short-lived when his hypercalcemia and symptoms returned. Elevated serum calcium and urine calcium levels in 1929; kidney stones in 1932; increasing bone pain and ultimately back in the hospital confined to a bed in 1933.[4] With obvious disease recurrence, Mandl decided to reoperate on Albert with encouragement from a presumed colleague, Julius Bauer.[5] Re-explore the neck with no preoperative imaging, unheard of in this day and age! This time Mandl left the operating room defeated. On October 16, 1933, all he found in Albert's neck was scar tissue; no local recurrence, no additional abnormal parathyroid despite exploration of the neck and mediastinum and removal of the thyroid[4]. Histologic examination revealed a small parathyroid of normal structure outside of the thyroid, an intrathyroidal parathyroid, and multiple microscopic foci of parathyroid tissue in the surrounding tissues.[4]

Did Dr. Mandl talk to his patient after the case? Did Albert just intuitively know that this time his hero was unsuccessful without any words exchanged? There was no change in Albert's symptoms or objective findings. Did Mandl make daily rounds to check on Albert? Did Albert remain hospitalized until his death from uremia on February 26, 1936? Was Mandl aware of Albert's death and autopsy findings? Autopsy pictured Albert with discordant femur length and multiple fractures, decalcified vertebral bodies, bilateral kidney stones and nephrocalcinosis, and pneumonia, yet no abnormal parathyroid tissue was present in the neck or chest (Figure 3). Unsatisfied with the autopsy findings, Niederle and colleagues made several attempts to secure the original histologic slides of Albert in the Museum of Medical History, but to no avail.

Did Albert succumb to metastatic parathyroid cancer? We continue to speculate that this was the case, though we will never know for sure. The known clinical data (severity of disease plus recurrence with lack of neck findings at reoperation) coupled with the gross and histologic findings in the original specimen (gray mass adherent to the recurrent laryngeal nerve with prominent fibrous bands) suggest the presence of parathyroid carcinoma.[7]

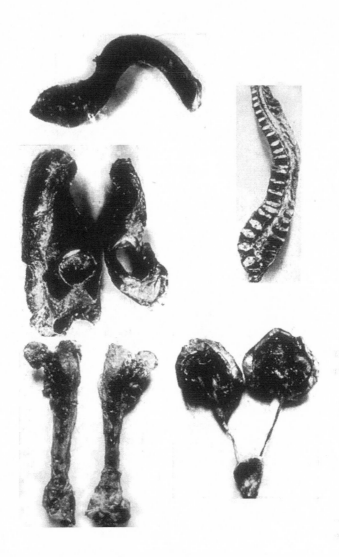

Figure 3: Autopsy specimens of Albert Jahne

In addition, Albert's initial postoperative course (lack of severe symptomatic hypocalcemia and early recurrence of kidney stones) favors persistent disease, possibly due to occult micrometastases. The postmortem examination did not include serial sectioning of distant organs for microscopic evaluation; therefore the presence of occult micrometastases cannot be definitively excluded.[4] The notion of parathyroid carcinoma in Albert Jahne is considered in Dr. Norman Thompson's writings on the subject[2] and supported by our personal communication with Dr. Aiden Carney.

Today, more than 80 years following Mandl's first successful parathyroidectomy for what the Viennese surgeon E. Gold later termed hyperparathyroidism,[4] we still reflect on his pioneering work, his courage and that of his patient. From the story of Mandl and Albert arose major tenets of parathyroid surgery that we continue to teach every day: removal of normal parathyroid tissue does not improve hyperparathyroidism, and parathyroid glands may be aberrant in position. In his 1940 publication in the *Journal of the International College of Surgeons,* Mandl estimated that some 300+ parathyroidectomies were performed from 1925 to 1940 with a mortality rate of 7-9%.[5] Now there are many times that figure performed annually with a nearly nonexistent mortality rate, thanks to the pioneering efforts of the generations of endocrine surgeons inspired by Mandl.

We continue to marvel at the trust that patients place in their surgeons' hands- this is as true today as it was in 1925. It is in so many ways a sacred covenant. The late Claude Organ reminded all of us, "an operation is an assault on a fellow human being- legalized but nonetheless an assault."

In his unique style, Alec Walt summarized this covenant, a covenant which we feel existed between Mandl and Albert and one which is present, as it should be, each and every time we take a patient to the operating room: "the concept that one citizen will lay himself horizontal and permit another to plunge a knife into them, take blood, give blood, rearrange internal structures at will, determine ultimate function, indeed sometimes life itself- that responsibility is awesome both in the true and in the currently debased meaning of the word." Albert Jahne was a hero, as are all patients everywhere.

References

1   Ellis H. The Cambridge illustrated history of surgery. 2nd ed.
     Cambridge; New York: Cambridge University Press, 2009.
2   Thompson NW. The history of hyperparathyroidism. *Acta Chir Scand*
     1990; 156(1):5-21.
3   Carney JA. The glandulae parathyroideae of Ivar Sandström.
     Contributions from two continents. *Am J Surg Pathol* 1996; 20(9):1123-
     44.
4   Niederle BE, Schmidt G, Organ CH, et al. Albert J and his surgeon: a
     historical reevaluation of the first parathyroidectomy. *J Am Coll Surg*
     2006; 202(1):181-90.
5   Mandl F. The Development of Parathyroidectomy during the Last
     Fifteen Years. *Journal of the International College of Surgeons* 1940; 3(4):297-
     311.
6   Delbridge LW, Palazzo FF. First parathyroid surgeon: Sir John Bland-
     Sutton and the parathyroids. *ANZ J Surg* 2007; 77(12):1058-61.
7   Sharretts JM, Kebebew E, Simonds WF. Parathyroid cancer. *Semin Oncol*
     2010; 37(6):580-90.

7

Captain Charles Martell
America's First Parathyroid Patient

Orlo H. Clark

At the 29th Annual Meeting of the Endocrine Society June 6, 1947, in his Presidential Address entitled "A Page of the History of Hyperparathyroidism," Dr. Fuller Albright compared the clinical course of the merchant sea captain Charles Martell and his battle with hyperparathyroidism to that of another Charles Martell who "rescued the Christian world from the clutches of the Saracens" in 732 at the Battle of Poitiers.[1] Although the French military victory may have had more immediate and dramatic results compared to the diagnosis of patients with primary hyperparathyroidism, the diagnosis and subsequent cure of patients with primary hyperparathyroidism has probably saved more lives.[1-5]

The patient Captain Charles Martell was born in 1889 and died in 1932 at the age of 43. He had been in excellent health until about age 22 when he developed symptoms that included groin, leg and hip pain. During the next three years he also developed pain in his heels, legs and back; he suffered fractures of his left clavicle, right humerus, and several vertebrae. By age 27 he became completely disabled and at age 30 on January 1926 he was hospitalized under the care of Dr. Eugene Dubois, who was Chief of Medicine at Bellevue Hospital in New York City and Professor of Medicine at Cornell University Medical School.[6] This was Captain Martell's fifth hospitalization for the myriad symptoms caused by an unknown condition, but it was his first hospitalization at an academic medical center.

By now, Captain Martell was an invalid and had lost seven inches in height. He had become pigeon-breasted, with marked kyphosis, muscular hypotonia, and weakness. (Figure 1).[7] Xrays revealed severe osteopenia, compressed vertebrae, as well as cystic changes in his femur (Figure 2).[8, 9] His blood calcium level was 14.8 mg/dl (normal range 8.5-10.5 mg/dL) with a phosphorus level 3.3mg/dl (normal range 2.5-4.5 mg/dL). Dubois was convinced that the captain had primary hyperparathyroidism, a previously unknown condition in the

Figure 1: A. Charles Martell, 10 years before his operations. B. Changes in skeleton due to Recklinghausen's disease of bone 10 years later.[29]

United States. In April 1927 he transferred Martell to the care of Dr. Joseph C. Aub, his former student and collaborator at Massachusetts General Hospital (MGH). Aub, who was funded to study lead poisoning, was also interested in calcium metabolism.[6] While Martell was at the MGH, Martell's blood calcium level ranged from 12-15 mg/dl with a blood phosphorus level ranging from 1.1-3.3 mg/dl. His calcium metabolism, as studied by Aub and his student Walter Bauer, was "similar to that of a normal man receiving 100 units of Collip's parathyroid extract daily."[1] Martell's diagnosis was confirmed in May 1927 by Aub, Albright, and Bauer.[10] Captain Martell was thus the first patient in the United States to be diagnosed with primary hyperparathyroidism, although not the first to be successfully treated.[6]

At the time of Martell's illness, parallel discoveries regarding the cause of primary hyperparathyroidism were being made in Europe. In 1925 Dr. Felix Mandl at the Hochenegg Clinic in Vienna had successfully diagnosed and operated on a 38 year-old streetcar driver, Albert Jahne, for osteitis fibrosa cystica and primary hyperparathyroidism. Jahne had clinical and metabolic findings similar to those of Captain Charles Martell.[11,12] Information relating to the diagnosis and successful treatment of Jahne was presented by Mandl at the Vienna Congress of Physicians on December 4, 1925, but this was apparently unknown by Dubois, Aub and colleagues in the United States.[13]

Figure 2: Hip. Upper right femur, showing largest bone cyst, and disappearance of cancellous and cortical bone.[40]

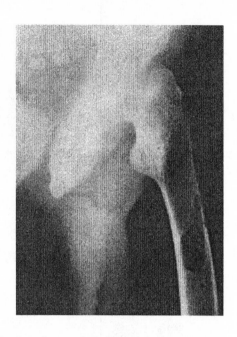

Discovery of the parathyroid glands and delineation of their function

What clinical and metabolic information was available to enable Mandl, Dubois, Aub and others to make the diagnosis of primary hyperparathyroidism and to have the courage to refer such patients for parathyroidectomy? We know that in 1850 Sir Richard Owen (1802-1892), the Hunterian Professor and Conservator of the Museum of the Royal College of Surgeons, identified a parathyroid gland while performing an autopsy of an Indian one-horned rhinoceros.[14] The parathyroid glands were first identified both grossly and microscopically in other animals and man by Ivar V. Sandstrom while he was a medical student in Uppsala, Sweden in 1879, but their function was unknown. Sandström's original, clearly written manuscript was rejected by the German press (probably because it was thirty pages long). Two abstracts regarding his findings, however, were published in German.[15,16] Parathyroid glands were probably also independently observed by Cresswell Baber (1851-1910) in England in 1881,[17] by Robert Remak (1815-1865) in 1855 in Berlin, and in 1863 by Rudolf Virchow (1821-1863), the founder of cellular pathology.[10] Remak was an anatomist and embryologist who described "the parathyroid gland as a small gland in association with the thymus that was clearly not a lymph node."[10] Virchow identified a "single specimen emphasizing that this was not an accessory thyroid gland, lymph node, or any other structure with which he was familiar."[10] The clinical importance of the parathyroid glands was not, however, recognized until the functional studies in 1892 by Eugene Gley (1857-1902).[17, 18] Gley's findings were confirmed by the Italian pathologists Guilio Vassale and Francesco Generali who documented that removal of the parathyroid glands, and not the thyroid gland, resulted in tetany.[19]

In 1907 William S. Halsted (1852-1922) wrote that "the remarkable discoveries by Gley and confirmed by Vassale and Generali of the vital importance of the parathyroid glandules stimulated afresh my interest in the surgery of the thyroid gland and suggested experiments in the transplantation of the tiny epithelial bodies."[20]

Parathyroid Pathology

In 1891 Professor von Recklinghausen (1833-1910) from Strasbourg, Germany at a Festschrift for his teacher Rudolf Virchow (1821-1902) "reported the autopsy findings, especially as related to the skeleton, in a series of 16 patients suffering from a variety of bone diseases."[1] Three of these patients had bone changes that represented the characteristic signs of hyperparathyroidism, now termed osteitis fibrosa cystica of von Recklinghausen.[1] These bony changes include "wide-spread fibrosis, cysts, and brown (or giant cell) tumors."[1] In 1903 Max Askanazy (1865-1940), a pathologist from Tubingen, Germany identified a parathyroid tumor in a patient with osteitis fibrosa cystica.[21] In 1906 Jacob Erdheim (1874-1937), a pathologist from Vienna reported that multiple enlarged parathyroid glands were present in patients with osteomalacia. Subsequently in 1914 he reported that similar parathyroid enlargement occurred in rats with spontaneous rickets.[22] He also confirmed Gley's findings and reported that "destroying the parathyroid glands with a cautery caused tetany and that young rats with chronic parathyroid insufficiency failed to calcify their teeth in a normal manner."[6]

Although many experts in the field at this time considered the enlarged parathyroid glands to be secondary to bone disorders, in 1915 Fredrich Schlagenhaufer (1866-1930), a pathologist from Vienna, reported that two patients with severe skeletal disorders had solitary parathyroid tumors.[23] Because of his findings he disagreed with the prevailing opinion that osteitis fibrosa cystica was the cause of the parathyroid tumor(s) and instead suggested that the parathyroid tumor(s) actually caused osteitis fibrosa cystica.[23] Evidence supporting Schlagenhaufer's theory was finally confirmed ten years later by Hoffheinz: "in von Recklinghausen's disease of bone the parathyroid enlargement (usually) involves only one gland."[24] In 1925 H. Hoffheinz[24] reported that among 44 patients with enlarged parathyroid glands found at autopsy, including one of his own patients "Osteitis fibrosa cystica occurred 17 times, osteomalacea (soft bones where mass of calcified bone is low) in 8, and rachitis or rickets (soft bone in childhood caused by vitamin D deficiency) twice."[24] Hoffheinz also documented that when parathyroid glands are enlarged in these patients, 85 percent had only one that was affected, "a finding difficult to explain on the basis of compensatory hypertrophy" where all parathyroid glands should be enlarged.[24] He astutely concluded that primary hyperparathyroidism was more likely to be the cause rather than a consequence of the abnormal calcium metabolism and skeletal disease.

Hyperparathyroidism and early research on calcium metabolism

In 1901 Jacques Loeb (1859-1924) at the University of California in Berkeley observed that increased neuromuscular irritability in response to a low blood calcium level in frog muscle could be reversed by administering calcium.[25] The observation suggested that calcium inhibited increased neuromuscular activity. These studies encouraged William G. MacCallum (1874-1944), a professor of pathology at Johns Hopkins Hospital in Baltimore, Maryland, to investigate whether a low blood calcium was responsible for the increased excitability of muscles observed in hypoparathyroid tetany. MacCallum and Carl Voegtlin (1879-1960) parathyroidectomized dogs and documented both that the blood calcium level decreased and that the administration of calcium promptly relieved the tetany.[26] Subsequent investigations on tetany by MacCallum and Vogel revealed conflicting results because they found that magnesium, and strontium as well as calcium relieved tetany.[27] They were also aware that Joseph and Meltzer had reported that treatment with hypertonic sodium chloride relieved post parathyroidectomy tetany.[28] MacCallum therefore thought that the cause of tetany was "some actual toxic material in the circulating fluids."[27, 29] Other studies by Leischer (1907) and Halsted (1909) documented that parathyroidectomy with subsequent removal of an autotransplanted parathyroid gland resulted in tetany thus providing important data regarding the function of the parathyroid glands.[30] Despite these classic experiments there was no generally accepted agreement regarding the function of the parathyroid glands.

Toxin theory

As previously mentioned, Vassale and Generali in 1900 confirmed Gley's reports; however, they suggested that the reason parathyroidectomy resulted in tetany was due to the fact that parathyroid glands had an antitoxic function.[31] They also reported that parathyroidectomy was less serious with fewer symptoms in old rather than younger dogs and more severe in dogs eating a meat diet. The symptoms in fasting dogs were less severe. These investigators were unaware of the effects of changes in blood pH levels and the importance of the blood ionized calcium levels relating to tetany.[31] "Arthur Biedl and others reported that copious bleeding and the transfusion of normal blood relieved the symptoms of tetany in parathyroidectomized animals."[31, 32] These observations therefore further supported their misconceived theory "that a toxin present in the blood is responsible for the symptoms and by the process of blood-letting the organism may be temporarily freed from the effects of the poison."[31, 32] MacCallum in 1905 and, Berkeley and Beebe in 1909, also removed blood in tetanic animals and found that the tetany resolved or became less severe.[33] In 1912-1913 W.F. Koch reported that methyl-guanidine and other bases were present in the urine of parathyroidectomized animals.[31, 34] Additional support of this observation occurred in 1924 when Noel Paton claimed that most evidence suggests "that complete removal of the parathyroid tissue leads to a fatal toxemia; he further suggests that the parathyroids through their internal secretion control the tone of muscles by regulating metabolism, i.e., the production and destruction of guanidine in the body."[31, 35] However, after reviewing all of the currently available evidence concerning the toxin theory, J.B. Collip (1892-1965), who was a professor of biochemistry in Edmonton, Alberta, concluded "that a very strong case has been made out for the guanidine intoxication theory of parathyroid tetany. The evidence, however, consists so largely of the circumstantial type that it fails to carry with it the weight of final conviction."[31] He documented that tetany could be successfully treated by injecting parathyroid hormone and that alterations of acid-base equilibrium influenced tetany in hypocalcemic animals.[31]

In 1923 Harold Salvesen, a Norwegian physician-scientist, also addressed the various theories regarding the function of the parathyroid glands.[36] At this time hypoparathyroid tetany was attributed to three possible conditions: hypocalcemia, guanidine intoxication, or to hypoglycemia. Salvesen convincingly demonstrated "that complete parathyroid ablation invariably lowered the blood calcium, that the blood sugar level was not altered and that guanidine accumulation occurred only terminally during agonal convulsions."[29] Salvesen's and Collip's findings convincingly established that post-parathyroidectomy tetany was related to low blood calcium levels and could be effectively treated with parathyroid hormone, thus refuting the toxin theory.

In 1924 Adolph Hanson (1888-1959) from Faribault, Minnesota and J.B. Collip independently extracted biologically active parathyroid hormone from the parathyroid glands of animals.[29] Hanson's preparation was used until the 1970s and named Parathyroid Extract USP. Collip's preparation was the purest and most potent parathyroid extract and was used in many important experiments in animals and man.[29] Injection of these parathyroid extracts caused hypercalcemia and hypophosphatemia.[37, 38]

The first successful parathyroidectomy worldwide

In 1924, Albert Jahne, a 34 year-old street car conductor in Vienna, Austria presented with a five year history of progressive weakness of his lower legs which interfered with his work.[12] Jahne was initially evaluated by Dr. Felix Mandl, who had been a pupil of Anton von Eiselsberg (1860-1939) professor of surgery in Utrecht and a student of Billroth. Jahne had bone pain in his legs and hips and marked fatigue. His previous bone x-rays a year earlier revealed not only transparent bones but also bone cysts in his pelvis and both femurs. Based on Erdheim's findings suggesting that the bone disease was the cause and not the consequence of an abnormal calcium level Mandl first treated Jahne with Collip's animal parathyroid extract.[6] It unfortunately did not help.[1] In December 1924 Jahne fell and fractured his leg and also developed "a white sediment in his urine." Still thinking that the bone disease was the cause of the problem, Mandl autotransplanted four human parathyroid glands from a fresh cadaver into Jahne but unfortunately there was no improvement. Therefore on July 30, 1925 Mandl explored Albert's neck and removed a 2.5x1.5x1.2 parathyroid tumor from the left lower position behind the thyroid gland. This left inferior parathyroid tumor was densely adherent to the recurrent laryngeal nerve. Three other normal parathyroid glands were also identified.[6] Within a few days Jahne's condition improved, and at about six days his urinary sediment disappeared. Surprisingly, he did not develop post-operative tetany since virtually all patients with osteitits fibrosa cystica experience bone hunger and profound hypocalcemia after parathyroidectomy. Jahne's blood and urinary calcium levels, however, decreased to normal.[7] After about four months his bone density improved and he was able to walk with crutches.[7]

The first reported parathyroidectomy in the United States

Unfortunately, information regarding the success of this operation in Vienna was apparently unknown by the physicians caring for Charles Martell in the United States.[13] Much of the other information regarding the function of the parathyroid glands was known, however: total parathyroidectomy resulted in tetany and animals treated with parathyroid extract developed hypercalcemia. Prior to the referral of Captain Martell to Dr Aub, Dubois and colleagues at Cornell Medical School and Bellevue Hospital, New York documented that Martell had hypercalcemia and intermittent hypophosphatemia.[9] Metabolic studies at the MGH confirmed these findings and also revealed that similar to Greenwald's dogs "which received an over abundance of Collip's parathyroid extract (100 units/daily) more calcium was going out than coming in" Captain Charles Martell.[1, 39] Dubois believed these metabolic studies confirmed the diagnosis of primary hyperparathyroidism.

Based on this information in May 1926, Elliot P. Richardson, Chief of Surgery at MGH and Professor at Harvard, did a unilateral neck exploration via a collar incision on Captain Martell and removed a normal right-sided parathyroid gland, "one of five nodules excised."[7] One month later Dr. Richardson explored Martell's left neck and a second normal parathyroid gland was excised with five other nodules. His caring physicians fooled themselves into thinking that Captain Martell improved clinically since Martell stated that he felt well and could "get about without difficulty" despite no appreciable change in his blood calcium level.[40] His bone x-rays surprisingly revealed improved calcification.[40] Although unable to work as a merchant Marine, Martell was able to work as a marine insurance broker. In 1929, three years after his first operations he was re-hospitalized at Cornell Medical Center and had a third negative neck exploration, this time by Russell Patterson. No abnormal parathyroid gland was identified. Unfortunately, Martell's symptoms became more severe, his renal function also deteriorated over the following 3 years and he was readmitted to the MGH in May of 1932. Fuller Albright, who had joined Aub and Bauer at the MGH in 1927, was now the chief of the metabolic clinic and had a major interest in calcium metabolism.[6] In 1931 Edward Churchill replaced Elliot Richardson as chief of surgery at the MGH. Churchill encouraged one of his surgical interns, Oliver Cope, to work with Benjamin Castleman, a pathology resident, to perform parathyroid dissections in 30 cadavers. Churchill was encouraged to do this study by Bauer and Albright, who had four patients awaiting parathyroidectomy.[13] Churchill and Cope subsequently performed several successful parathyroidectomies in 1932. At the conclusion of the first successful parathyroidectomy, Churchill told Cope that he was "too tough, and that the field was too bloody."[13] They also re-operated on Captain Martell's neck three times, but without success. Martell was not ready to give up, apparently having read extensively about his own condition in the Harvard Medical Library. Based upon his own research, he advised his physicians to perform a mediastinal exploration.

At Captain Martell's seventh parathyroid and first mediastinal exploration, Drs. Churchill and Cope performed a subtotal resection of a 3x3 cm mediastinal parathyroid adenoma, situated just lateral to the superior vena cava, leaving about ten percent of the vascularized parathyroid remnant at the suprasternal notch in an attempt to decrease the risk of postoperative tetany.[10] Unfortunately, tetany developed on the third postoperative day and Martell was treated with calcium and Collip's parathyroid extract. Six weeks after this successful operation he developed ureteral obstruction and unfortunately died of laryngeal spasm after a urologic procedure to relieve his obstruction.[10]

Although today many medical journals no longer accept case reports for publication, Captain Martell's case history dramatically advanced the care of patients with primary hyperparathyroidism. Captain Martell was extensively studied in metabolic units at both Cornell and the MGH. His battles with hyperparathyroidism were closely catalogued, leading to seven case reports and two presidential addresses.[6] The collaborative efforts of the endocrinologists, surgeons, and pathologists at the MGH led to their ultimate leadership in the diagnosis and treatment of patients with hyperparathyroidism that continues to this day.

References

1   Albright F. A page out of the history of hyperparathyroidism. *J Clin Endocrinol Metab* 1948; 8(8):637-57.

2   Clark OH. "Asymptomatic" primary hyperparathyroidism: is parathyroidectomy indicated? *Surgery* 1994; 116(6):947-53.

3   Rienhoff WF. The surgical treatment of hyperparathyroidism, with a report of 27 cases. *Ann Surg* 1950; 131(6):917-44.

4   Hellstrom J. Further observations regarding the prognosis and diagnosis in hyperparathyroidism. *Acta Chir Scand* 1953; 105(1-4):122-31.

5   Hedbäck G, Tisell LE, Bengtsson BA, et al. Premature death in patients operated on for primary hyperparathyroidism. *World J Surg* 1990;

14(6):829-35; discussion 836.

6   Thompson NW. The history of hyperparathyroidism. *Acta Chir Scand* 1990; 156(1):5-21.

7   Thomas CG. The glands of Owen--a perspective on the history of hyperparathyroidism. *Surgery* 1990; 108(6):939-50.

8   Bauer W, Albright F, Aub JC. A Case Of Osteitis Fibrosa Cystica (Osteomalacia?) With Evidence Of Hyperactivity Of The Para-Thyroid Bodies. Metabolic Study Ii. *J Clin Invest* 1930; 8(2):229-48.

9   Hannon RR, Shorr E, McClellan WS, et al. A Case Of Osteitis Fibrosa Cystica (Osteomalacia?) With Evidence Of Hyperactivity Of The Para-Thyroid Bodies. Metabolic Study I. *J Clin Invest* 1930; 8(2):215-27.

10 Organ CH. The history of parathyroid surgery, 1850-1996: the Excelsior Surgical Society 1998 Edward D Churchill Lecture. *J Am Coll Surg* 2000; 191(3):284-99.

11 Mandle F. Therapeutischer versuch bein einem falls von otitis fibrosa generalisata mittles. Exstirpation eines epithelkorperchentumors. *Wien Klin Wochenschr Zentral* 1926; 53:260-264.

12 Niederle BE, Schmidt G, Organ CH, et al. Albert J and his surgeon: a historical reevaluation of the first parathyroidectomy. *J Am Coll Surg* 2006; 202(1):181-90.

13 Cope O. The study of hyperparathyroidism at the Massachusetts General Hospital. *N Engl J Med* 1966; 274(21):1174-82.

14 Felger EA, Zeiger MA. The death of an Indian Rhinoceros. *World J Surg* 2010; 34(8):1805-10.

15 Sandstrom I. Glandulae parathyroideoe (abstract). *Schmidt's Jahrbuch d ges.* 1880(187):114-118.

16 Sandstrom I. Glandulae parathyroideoe (abstract): Jahresber Fortschr Anat. *Physiol* 1880; 9:224-226.

17 Welbourn RB. The history of endocrine surgery. New York: Praeger, 1990.

18 Gley E. Functions of the thyroid gland. *Lancet* 1892:142.

19 Vassale G, Generali F. Fonction parathyréoidienne et fonction thyréoidienne. *Arch ital de beol* 1896; 33:154-156.

20 Halsted WS. Hypoparathyreosis, status parathyreoprivus and transplantation of the parathyroid glands. *A J Med Sci* 1907; 134:1-12.

21 Askanazy M. Uber ostitis deforms ohre osteides Gewebe. *Arb path Anat. Inst Tubingen (Leipzig)* 1904; 21:398-422.

22 Erdheim J. Rachitis and Epithelkorperbefudnl: Kaiserlich-Koniglichen

Hof-und Staats druckerie, Wien 1914.

23 Schlagenhaufer F. Zwei falle von parathyroideatumoren. *Wien Klin Wochenschr* 1915; 28(2):1362-end.

24 Hoffheinz H. Über Vergrößerungen der Epithelkörperchen bei ostitis fibrosa und verwandten krankheitsbildern. *Virchows Arch Pathol Anat Physiol Klin Med.* 1925; 256(3):705-35.

25 Loeb J. On the different effects of ions upon myogenic and neurogenic rhythmical contractions and upon embryonic and muscular tissue. *Am J. Physiol* 1899-1900; 3:383-87.

26 Maccallum WG, Voegtlin C. On The Relation Of Tetany To The Parathyroid Glands And To Calcium Metabolism. *J Exp Med* 1909; 11(1):118-51.

27 Maccallum WG, Vogel KM. Further Experimental Studies In Tetany. *J Exp Med* 1913; 18(6):618-50.

28 Joseph DD, Meltzer SJ. The inhibitory action of sodium chloride upon the phenomena following the removal of the parathyroids in dogs. *JPET* 1911; 2(4):361-374.

29 Goldman L, Gordan GS, Roof BS. The parathyroids: progress, problems and practice. *Curr Probl Surg* 1971:1-64.

30 Garrison FH. An Introduction to the History of Medicine. Fourth Edition ed. Philadelphia: W.B. Saunders, 1960.

31 Collip JP. The parathyorid glands. *Medicine* 1926; 5:1-57.

32 Biedl A. The internal secretory organs: their physiology and pathology. New York: William Wood and Company, 1913.

33 Berkeley WN, Beebe SP. A Contribution to the Physiology and Chemistry of the Parathyroid Gland. *J Med Res* 1909; 20(2):149-73.

34 Koch WF. On the occurrence of methyl guanidine in the urine of parathyroidectomized animals. *J Biol Chem* 1912; 12(3):313-5.

35 Paton N. Recent investigations on tetania parathyreopriva and idiopathic tetany, and on the functions of the parathyreoids. *Edinburgh Med J* 1924; 31(10):541-59.

36 Salvesen HA. Studies on the physiology of the parathyroids. *Acta Med Scand* 1923; 6(suppl):1-160.

37 Hanson A. A parathyroid preparation for intramuscular injection. *Military Surgeon* 1924; 54:218-9.

38 Collip JB. The extraction of a parathyroid hormone which will prevent or control parathyroid tetany and which regulates the level of blood calcium. *J Biol Chem* 1925; 63:395-438.

39 Greenwald I, Gross JJ. The effect of long continued administration of parathyroid extract upon the excretion of phosphorus and calcium. *J Biol Chem* 1926; 68:325.

40 Richardson EP, Aub JC, Bauer W. Parathyroidectomy In Osteomalacia. *Ann Surg* 1929; 90(4):730-41.

8

First Encounters with Pheochromocytoma
The Story of Mother Joachim

Jon A. van Heerden

Although there had been prior scattered reports in the literature about paroxysmal hypertension, in 1926 pheochromocytoma and its relationship to hypertension were unknown. Catecholamines were first isolated in 1898 by Abel, but it was not until 1929 that Rabin first demonstrated increased epinephrine content in the adrenal medulla.[1] Roux in Lausanne, Switzerland performed the first successful resection of a pheochromocytoma in February 1926. The case of Mother Joachim represents, to the best of our knowledge, the first successful resection of a pheochromocytoma after Roux's patient and certainly the first in the United States.

Mother Mary Joachim was a 30-year old Roman Catholic nun from Chattam, Ontario, Canada (Figure 1). She was referred to the Mayo Clinic by J.H. Duncan of Chattam, who wrote in his referring letter, dated May 31, 1926: "I feel much as Festus felt in sending Paul to Rome – not having any definite accusation against him... She complains of weakness and also complains and very much fears recurrent attacks of gas, nausea and vomiting associated with severe headaches... I believe that these attacks are due to an incompetent liver." Dr. Duncan concluded, "I suggested to Mother that with your extensive and specialized machinery for diagnosis, you might readily find some pathological condition that I have missed."

Mother Joachim arrived at the Mayo Clinic on June 3, 1926, and was admitted to Saint Mary's Hospital the following day for investigation and observation. During this period, which lasted until the operation on October 11, many astute observations were made by both the attending physicians and the nursing staff. Daily recording of her blood pressure revealed levels of 100/70 to 280/190 mm Hg. The descriptions of her attacks were summarized and accurately recorded on a number of occasions:[1] "Sudden onset – pain in back – some pain occipital region and right side of neck associated with nausea and vomiting. Cold and clammy, rapid respirations, rapid heartbeat."[2]

Figure 1: Mother Mary Joachim

"First note a slight anxious expression to face. Pupils slightly dilated. Pulse 124, full and pounding with radial vessels tight and easily rolled. Blood pressure, 300/160 mm Hg after 2 minutes. Patient felt nauseated. Some vasomotor mottling about mouth and nose. Aorta palpable in episternal notch, neck veins full. Heart sounds pounding. Hands cool and sweaty. No loss of consciousness."[3] "Sudden onset of feeling of suffocation with nausea and vomiting. Blood pressure, 260/190 mm Hg. Pulse rate, 144 per minute. Face intensely cyanosed. Respiration labored and rapid – 48 per minute."

L.G. Rowntree, an internist, believed that the hypertension was mediated through the sympathetics and approached Dr. Alfred Adson to consider cervical sympathectomy. C.H. Mayo who was asked to see the patient in September, stated: "Toxins evidently intermittently discharged affecting the sympathetic." He also thought, however, that the patient's complaints were largely psychosomatic.

During Mother Joachim's prolonged period of preoperative observation, many drugs were tried in an effort to control her attacks, including quinidine, morphine, veronal, sodium nitrate, digitalis, atropine, chloral hydrate, sodium bromide, potassium iodide, phenobarbital, histamine, tincture of belladonna, and amyl nitrate.

Because of persistent lumbar pain, a surgical exploration was performed by Dr. Mayo in October 11, 1926. The first assistant was Claude Dixon, who later (in 1928) joined the general surgical staff and gained recognition in his own right for his surgical dexterity. The operation lasted from 8:10 to 9:14 a.m. According to the operative dictation: "Exploration of the left adrenal showed it to be twice normal size. Rounded tumor apparently the size of

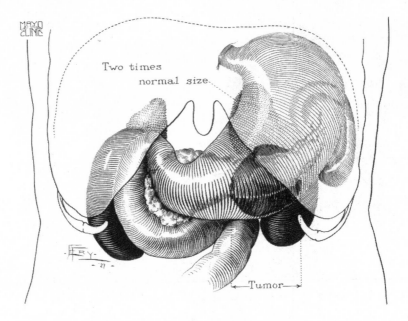

Figure 2: Drawing of left-sided pheochromocytoma and anatomic relationships

a lemon was situated beneath the tail of the pancreas and the inner side of the left kidney. Opened the outer capsule of the retroperitoneal tumor and enucleated it intact. Considerable venous hemorrhage which was controlled by pack. On sectioning the tumor, it looked like a brain tumor." The pathologic interpretation by W.C. MacCarty was as follows: "Encapsulated fibrous cellular retroperitoneal neoplasm. Malignant." The anatomic relationships of the tumor are shown in Figure 2.

Postoperatively, the patient's blood pressure never increased above 130 mm Hg. and except for a basilar pulmonary infiltration her postoperative course was uneventful. She was finally discharged and returned home on December 13, 1926.

The day after the operation (October 12), Dr. Mayo wrote to Dr. Duncan as follows: "Yesterday I operated on your patient Mother Joachim, exploring the abdomen and finding an encapsulated tumor on the left side of the spine over the kidney; I removed the tumor which was about 1 ½" thick and about 2 ½" across. It is evidently malignant, but the tissue looks like nerve tissue. Her blood pressure immediately dropped and she feels much better this morning than she has for a long time."

In a letter to Dr. Mayo dated February 28, 1927, Mother Joachim wrote:

"I am feeling wonderfully well and gaining weight each day. I have never had that terrible pain in my head again." In a further letter dated March, 1927, she wrote: "I have gained 30 pounds since I came home – that was 11 weeks ago. I have never weighed so much in my life. Isn't that good?" And in July 1927: "At present, I am anxious to practice on the violin for about an hour a day. Do you think the exertion would be in any way injurious? Reverend Mother is grateful for my present health. Therefore, she does not want me to undertake any work that might retard my present health and permanent recovery, and would like your advice in this matter." In January 1931: "It is a great pleasure for me to write you that I am enjoying excellent health. I have not had a sick spell since I left Rochester."

Dr. Mayo presented a case report on Mother Joachim before the Section of Obstetrics, Gynecology and Abdominal Surgery at the 78th annual session of the American Medical Association that was held in Washington, D.C. on May 20, 1927.[2] In his comments, Dr. Mayo postulated:[1] "Among other possible causes for the hypertension, irritation of the abdominal sympathetic by the tumor must be considered," and[2] "Epinephrine acts quickly on arterial and capillary vessels, and the suprarenal probably have much to do with quick vascular changes in this region of the abdomen, which is almost a reservoir for blood; it is here that many temporary changes probably occurred in surgical shock and collapse."

This report prompted a letter from Harvey Cushing of the Peter Bent Brigham Hospital dated May 21, 1930:

Dear Charlie: You will, of course, remember that remarkable case you reported two years ago in the Journal of the AMA on paroxysmal hypertension with recovery following your brilliant operation disclosing a retroperitoneal tumor that looked like a tumor of the adrenal medulla – probably, if I remember correctly, due to a tumor of the chromaffin system elsewhere.

I find that most tumors of this kind are highly malignant and I am anxious to know what has been the later history of this woman's case. Should the growth have recurred one would naturally expect a recurrence of her former symptoms of hypertension. I hope, for the patient's sake that this hasn't happened, but it would be a matter of great interest to me, as I am puzzling my mind over some hypertensions now under observation here.

Dr. Mayo replied:

Dear Cushing: I am enclosing a reprint concerning the case of paroxysmal hypertension with tumor of retroperitoneal never. The patient is still doing fine. The tumor was apparently an enormous enlargement of one of the ganglion which rested on on the renal vessels close to the kidney and the blood supply of this area was enormous – I had to pack the wound to control it.

In a letter to Dr. Claude Dixon dated March 17, 1934, Mother Joachim wrote: "Quite recently, I suffered the loss of my dear father whose years had reached four score and one. As my health is now, I see no reason why I cannot equal his record." Dr. Mayo replied to her letter (June 21, 1934) and included a reference to his own health, which was declining at that time.

I am pleased to be able to tell you that I am improving every day. My health was a little down for a couple of months. However, I am back in the harness, a little lighter one than I used to wear, and I go to the operating rooms in the morning and see patients in the Clinic in the afternoon.

On October 13, 1941, she wrote to another Mayo Clinic physician (Dr. Mayo had died in 1938):

Recently, I received your letter asking me to return to Rochester for a checkup at the hospital. I referred the matter to Reverend Mother and after careful consideration of the present day conditions, she finds it quite impossible to send any nuns to Rochester. There is great difficulty in obtaining passports and American money. Reverend Mother wishes to thank you for the interest you have taken in me and is most grateful to you and the Mayo Clinic for all the kindness shown to the Ursulines when we were in Rochester. I am enjoying good health, I have a large music class and am able to do a full day's work.

Mother Joachim died in May 1944, at the age of 48, 18 years after her operation. Notice of her death was received in a letter from Mother M. Gertrude:

Mother Mary Joachim's operation at the Mayo Clinic was, we feel, entirely successful, and there is not doubt of its prolonging her life many years. However, she seems to have had a heart condition of which she said very little. Coronary thrombosis took her while she slept. She had taught her music pupils the previous day and was cheery and bright as usual.

It has been written that the Mayo brothers "made Listerian surgery almost as reliable a science as bookkeeping.[3] The story of Mother Joachim demonstrates how astute observations and accurate recoding of the then-unknown paved the way for those who followed and permitted elucidation and identification of this rare but fascinating endocrine tumor, pheochromocytoma.

A previous version of this chapter was published in the *American Journal of Surgery* (1982 Aug;144(2):277-9). Reprinted with permission.

References

1   Berglind RM, Harrison TS. Pituitary and adrenal. 3d ed. New York: McGraw-Hill, 1979.

2   Beahrs OH. American Association of Endocrine Surgeons. Presidential address: Lest we forget. *Surgery* 1987; 102(6):893-7.

3   Garrison FH. An introduction to the history of medicine, with medical chronology, suggestions for study and bibliographic data. 2d ed. Philadelphia,: Saunders, 1917.

9

Injury to the Superior Laryngeal Branch of the Vagus Nerve During Thyroidectomy
Lesson or Myth?

Peter F. Crookes

Introduction

Many younger surgeons receive their earliest exposure to surgical history while assisting a senior surgeon with an interest in the disease or the procedure in question. Across the operating table stories tend to be shared in the hope that the human and historical elements of the case will fix some important principle in the mind of the trainee. It is rare for these stories to be checked for historical accuracy. Occasionally a good story illustrating an important point of management or technique comes to be enshrined in textbooks and articles, and is perpetuated despite clear evidence to the contrary. The widespread belief that a famous opera singer lost her voice, causing a premature and precipitous end to a glorious career, as a consequence of a thyroidectomy is one such story, and I intend to show in this chapter that the belief is unfounded.

Recurrent nerve injury during thyroidectomy is the best-known and best-studied complication of this operation. The external branches of the superior laryngeal nerve may also be injured as the superior thyroid pedicles are ligated – they are smaller, encountered at the extreme of the field of dissection, and their anatomic course is not as familiar to surgeons. Techniques for routinely protecting the superior laryngeal nerves are not commonly found in textbooks of operative surgery. Consequently there is widespread lack of awareness of their anatomic proximity and the effects of injury. It is believed that superior laryngeal nerve injury alters the speaking voice relatively little, but that its effects are more obvious when singing, especially high notes, and the singer can no longer generate the required tension in the cricothyroid muscle. To illustrate these effects, many textbooks and articles quote the case of the famous operatic soprano Amelita Galli-Curci (1882-1963) as an example of the

Figure 1: Amelita Galli-Curci
Courtesy New York Public Library

disastrous effects of superior laryngeal nerve injury on vocal performance.[1-3] The purpose of the present article is to examine the evidence that superior laryngeal nerve damage during thyroidectomy was the cause of Galli-Curci's vocal decline in the years following her operation.

The patient

Amelita Galli was born in Milan in 1882, though as an adult she always gave her birth year as 1889. She came from a musical family and studied piano at the Milan conservatory of music, where she graduated with a gold medal in piano in 1905. She received almost no formal vocal training, but she was encouraged to seek a career as a singer by the composer Pietro Mascagni (1863-1945), who heard her sing during a visit to her family home. She made her operatic debut in Trenta, Italy, in 1906, and over the next few years she toured extensively in Italy, Spain, Egypt and South America to great acclaim. In November 1916, as a relatively unknown singer in the United States, she was allocated the lead role in the Chicago Opera singing Gilda in Verdi's *Rigoletto*, creating a spectacular sensation that launched her American career as a coloratura soprano, where she was a star attraction every season in Chicago and New York for the next 14 years. In the late 1920s critics began to notice a decline in her vocal powers, commenting on her tendency to sing flat, and loss of the sense of effortless ease that previously characterized her execution of high and difficult passages. She retired from opera singing in 1930. Thereafter she went on several long tours as a recital singer, always singing to packed houses in Britain, continental Europe, Australia, India and the Far East.

In the early months of 1935 while on tour in India she noticed more obvious vocal difficulty, attributing it to the dry and dusty air. A chance meeting in the hotel in Rawalpindi, when Galli-Curci's second husband Homer Samuels met a patient of his twin brother's, brought to light the fact that an American surgeon with a special interest in goiter was staying in Shrinigar studying the effects of dietary deficiency of iodine. This surgeon was Arnold Kegel. Galli-Curci had known him in Chicago, and immediately telegraphed him. He traveled over 200 miles by car to Rawalpindi to see her and examine her the next day. She was found to have a substantial goiter and, when examined by indirect laryngoscopy, considerable tracheal compression. It emerged that Galli-Curci had been aware of the goiter for several years and had increasingly noticed a sense of constriction as it enlarged. Early photographs show no trace of it, but it is clearly visible in later photographs (Figures 1-3). Dr. Kegel recommended further study, and he subsequently met her again in Calcutta, and traveled with her to Rangoon, Singapore and Java, finally ending up at the University Hospital, Tokyo, where he was able to obtain a more exact radiographic study. These studies revealed that the larynx was displaced 1.5 inches to the left side by the goiter and the tracheal diameter was reduced by 50%. At this point, he recommended thyroidectomy. He arranged to re-examine her when she returned to the United States.

The surgeon

Arnold Henry Kegel was born in Lansing, Iowa, in 1894, and studied medicine at the University of Illinois, where he graduated with an MD in 1916. He did surgical training at the Mayo Clinic between 1917 and 1921. He subsequently returned to Chicago where he established a busy practice, frequently performing thyroidectomy using a method requiring only local anesthesia so that he could make the patient vocalize during surgery to ensure that he was not endangering the laryngeal nerves. The years around 1930 were notable for the frequency of thyroidectomy, and it is estimated that at the Mayo clinic alone

Figure 2: Galli-Curci in 1919, showing no evidence of thyroid enlargement. Courtesy New York Public Library

there were 3000 thyroidectomies carried out annually.[4] In 1927 he was asked by the Mayor of Chicago to become the Commissioner of Health, apparently as a result of his ability to sort out a hospital plumbing problem that had led to an outbreak of cholera. Between 1928-1932, he apparently did little or no surgery, but devoted himself to public health issues such as the risks of domestic refrigerators and the merits of compulsory vaccination of schoolchildren.[5, 6] He played a prominent role in the famous "baby-switching" case in Chicago in 1930, by using the relatively new techniques of blood typing to establish the claim by one set of parents that their baby had been given to another couple in error.[7] He left Chicago shortly after returning to practice and moved to Los Angeles in 1935, and before he was eligible for a California license he took a trip to North India, where he came into contact with Galli-Curci.

In later years Dr. Kegel continued to perform thyroidectomy under local anesthesia, certainly up until the 1950s (Kenneth Morgan, MD, personal communication), but progressively concentrated on the non-operative approaches to gynecologic problems, namely urinary incontinence and sexual dysfunction. He invented what may have been the first example of biofeedback, which he used clinically to develop improved tone and awareness in the pelvic muscu-

Figure 3: Galli-Curci after developing a visible goiter. Courtesy New York Public Library

lature in women.[8] His name is now chiefly associated with exercises designed to improve urinary incontinence and sexual dysfunction by developing the pubococcygeus muscle. He died of an aortic aneurysm in 1981.

The operation and its sequelae

When Galli-Curci returned to the United States, Dr. Kegel arranged to perform a thyroidectomy but had to admit the patient to Henrotin Hospital in Chicago since he still did not have a California license. Thyroidectomy was performed on August 11, 1935. The assistant was G. Raphael Dunleavy MD, a Los Angeles surgeon who had graduated from Northwestern University School of Medicine and trained in surgery at Los Angeles County Hospital. He was also evidently Dr. Kegel's brother-in-law. The operation was performed under local anesthesia, and at several points the surgeons asked the patient to sing scales to ensure that the laryngeal nerves were not traumatized. At the end, she sang part of a duet from *The Barber of Seville*. An artist, Ms. Lucy Bassoe, was present in the operating room but the sketches and the operative record have been lost. The patient's first vocal exercises were performed in the ward, and her

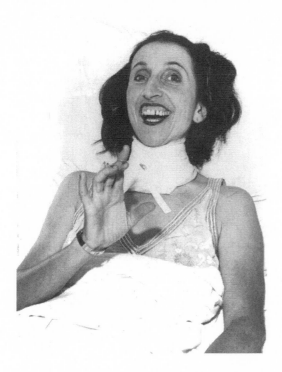

Figure 4: Galli-Curci recovering after her thyroidectomy in 1935. Courtesy New York Public Library

voice was initially harsh. When one of the nurses commented "Wonderful, Madame" she replied acerbically "Wonderful? It sounds like a buzz saw hitting a rusty nail!" There were no postoperative complications. (Figure 4)

She was discharged from the hospital on August 18 and returned to Los Angeles by rail, accompanied by Dr. Kegel. Even before discharge, she was euphoric about the change in her voice. She attributed the improvement to the facilitation of air flow which followed the removal of what she described as her "little potato in the throat."[9] She undertook more formal re-training from her former brother-in-law Gennari Curci. The director of the Chicago City Opera Company was contacted and informed about the "new voice." The opera company was at that time in some difficulty and jumped at the chance to regain their audiences by announcing the return of Galli-Curci to the opera stage, though she had not appeared in an opera for six years and was now in her mid-fifties. News that she was to sing opera again was a great stimulus for the opera company and subscription seats were rapidly sold. On November 16, 1936, 20 years to the day since her first spectacular debut as an opera singer in the U.S., she appeared in Chicago as Mimi in *La Bohème*, being greeted with prolonged and tumultuous applause at her first entrance. Unfortunately, her performance was harshly reviewed by many critics. Some critics withheld judgment, but others dubbed the performance "pathetic."

Despite this isolated failure, she subsequently made several broadcasts later in the year, and sang in several recitals throughout 1937. In contrast with the years 1920-30, very few reviews of her public performances after her surgery survive. She gave an acclaimed recital in Albany on April 10, 1937, and did a Pacific Coast tour from Los Angeles to Victoria BC in the fall. Press reviews from Los Angeles, Seattle, Victoria and Winnipeg were collected and included in her biography,[10] but for the year 1937, the only review preserved in the archives of the New York Library for Performing Arts records a performance in Wadena, Minnesota, a town of less than 5000 people. It is unlikely that a town of that size possessed an opera critic of sufficient discriminatory ability to assess her vocal capacities. Similarly, the conductor of the Detroit Symphony Orchestra, who accompanied her in a live broadcast in December 1936 said "she sang beautifully tonight" but had never heard her in her prime. In 1938, she abruptly cancelled her engagements and retired from singing. She continued to live in Westwood, California until 1945. During this time she maintained frequent social contact with Dr. Kegel, who lived in the same neighborhood. Thereafter she retired to Rancho Santa Fe, California, and lived a relatively secluded life, devoting herself to painting and piano playing. She was friendly with Paramahansa Yogananda, head of the Self Realization Fellowship in Encinitas, CA but had few close companions, especially after her husband died in 1954. In her eighties she built a smaller house in nearby La Jolla where she moved in 1962. She died on November 24, 1963, apparently from emphysema.

Discussion

The anatomy and physiology of the superior laryngeal nerves are now quite well-understood and several anatomists and surgeons have systematically dissected the nerves in an attempt to define how vulnerable they are during thyroidectomy. It is estimated that the course of the external branch of the superior laryngeal nerve lies immediately adjacent to the superior thyroid artery in 21% of cases, and is thus at high risk of being injured during mobilization of the superior pole of the thyroid.[11] Though these details are of fairly recent description, injury to the superior laryngeal nerve was known in the 1920s and the relevant literature is likely to have been known to Dr. Kegel.[12] The operative notes relating to Galli-Curci's thyroidectomy have been lost and the hospital has been demolished. There is consequently no direct evidence as to whether Dr. Kegel was aware of the nerves or made any attempt to protect them. The typical technique for ligating the superior pole vessels was blind ligation by passing a ligature threaded on an aneurysm needle. However, there are several reasons why it seems unlikely that the thyroidectomy performed by Dr. Kegel caused injury to the superior laryngeal nerve.

Firstly, subtle vocal changes in timbre and agility are usually noticed by the singer in private even when they are not obvious to listeners. A noted laryngologist with extensive experience in vocal cord injuries in singers has observed that the singer is always the first person to notice the problem, often long before it is detectable to listeners (Herbert Dedo, MD, personal communication). It seems unlikely that Galli-Curci, conscious of her declining vocal powers in the years before her surgery, could have suffered an injury to her laryngeal nerves of which she herself was unaware. She was sufficiently self-critical for the initial postoperative difficulty to be obvious to her. It is undeniable that in the year before she died she gave an interview in which she attributed the end of her singing career to her operation, but this is at variance with her enthusiastic comments in the early postoperative period and throughout 1937. Further, the late Joan Sutherland spent an engaging day with her in 1961 and they evidently discussed all sorts of matters relating to singing technique and artistry. In my personal conversation with Joan Sutherland about the event, she was aware that Galli-Curci had undergone a thyroid resection, but she told me that Galli-Curci made no mention of the effect of her operation on her career. Rather, it is likely that the improved air flow following the removal of the goiter imparted a sense of freedom which allowed her for a time to avoid confronting the fact that her voice had been declining steadily for several years. Evidence of this deterioration is now readily available, as her later recordings have been recently compiled in CD form. A comparison of her 1917 and 1930 recordings of the aria *Caro Nome* from Rigoletto is instructive. Despite the superior technical quality of the later recording, the vocal performance shows unmistakable signs of deterioration, especially in the inability to negotiate the rapid decorative passages towards the end, and the sustained penultimate note (high B) is almost a semitone flat.

Secondly, Galli-Curci remained on friendly social terms with Dr. Kegel for several years afterwards. Dr. Kegel's son has told me that he and his father frequently visited the home of Galli-Curci and her husband until the 1940s. This behavior seems out of keeping if she believed that the operation had prematurely aborted her career. Filing a lawsuit would be a much more typical response, and Galli-Curci was no stranger to the courts in this country, having sued her first husband over a dispute about his right to a share in her vast earnings.

Thirdly, the pattern of vocal change is inconsistent with what is now known of the effects of superior laryngeal nerve injury. These include a shallow voice, fatigue of the voice, and a tendency to aspiration. Spontaneous recovery, either by genuine resolution of nerve pathology or unconscious compensation by other vocal cord elements, may occur within six months.[13]

Fourthly, the gradual vocal decline which affects most female singers especially at the extremity of their range typically begins in their forties and may be associated with menopausal changes in vocal cord morphology.[14-16] An additional possible factor in Galli-Curci's case is the well-known effect of thyroid deficiency on the voice, and though it is not possible to know whether she required or took thyroid supplements, it is known that she gained 15 lb in the year after the operation. In any case, by the age of 50 most operatic sopranos give up these coloratura roles, typically depicting young heroines, and assume lyric soprano roles more appropriate to their stage in life. Very few sopranos have continued to sing these coloratura roles after the age of 50 (Table 1). Galli-Curci later remarked that the true coloratura's career begins to decline after age 35. Recent support for this view comes from a notable present day heir of Galli-Curci, the Korean coloratura soprano Sumi Jo. In a broadcast interview on KUSC on December 5, 1999, she indicated that she no longer sings the famous "Queen of the Night" aria from Mozart's *Magic Flute*, even though she is still in her thirties.

Galli-Curci's disappointing performance at her "second debut" has thus more obvious explanations. Although her towering reputation during the 1930s was maintained by her success in the less demanding role of a recital singer, no more severe test could be imagined than for her to appear at the age of 54 as a nervous and corpulent middle-aged woman who had not sung on the opera stage for over six years, to play the part of a young consumptive heroine.

It is thus natural to wonder how the story originated. It appears to have been promulgated by the late Timothy Harrison MD (1927-2010), a famous endocrine surgeon from University of Michigan, and later Hershey, PA, who wrote the chapters on surgery of the thyroid gland in several successive editions of Sabiston's Textbook of Surgery. I had the opportunity to discuss his views with him, but he adduced no further evidence than his personal conviction and the apparent temporal relationship between the surgery and the vocal decline. Logically this may constitute the *post hoc ergo propter hoc* fallacy. It nevertheless remained sufficiently influential to reappear in a recent textbook on thyroid surgery.[17]

In light of these considerations, the evidence that Dr. Kegel's thyroidectomy precipitated the end of Galli-Curci's career as a soprano evaporates. It has a superficial plausibility and may help emphasize to the surgeon-in-training how important it is to protect the superior laryngeal nerves, and to remind physicians of their importance in vocal function. However, in the interest of historical accuracy and in fairness to the reputation of the surgeon concerned, the story should be relegated to the realm of myth and henceforth expunged from future surgical textbooks and articles.

Acknowledgements

I gratefully acknowledge the help and information given by phone and letter from Mr. Robert A Kegel, (Dr. Kegel's son), Edward Morgan, MD, my colleague Gail Anderson MD, and Ms. Frances Wright RN, all of whom supplied personal reminiscences of Dr. Kegel. Frank Dedo MD shared his extensive experience of his close association with the San Francisco Opera, and his studies in laryngeal nerve injury. Galli-Curci's discographer, Mr. Bill Park, and the biographical information on the website of Mr. John Craton (craton@ youpy.fr) were also very helpful. The staff of the New York Public Library for the Performing Arts kindly placed all their archival material at my disposal and provided the hitherto unpublished photographs.

A previous version of this chapter was published in *Annals of Surgery* (2001 Apr;233(4):588-93). Reprinted with permission.

References

1   Kark AE, Kissin MW, Auerbach R, et al. Voice changes after thyroidectomy: role of the external laryngeal nerve. Br Med J (Clin Res Ed) 1984; 289(6456):1412-5.

2   Choksy SA, Nicholson ML. Prevention of voice change in singers undergoing thyroidectomy by using a nerve stimulator to identify the external laryngeal nerve. Br J Surg 1996; 83(8):1131-2.

3   Gulec SA, O'Leary JP. Fable on the superior laryngeal nerve. Am Surg 1999; 65(5):490-2.

4   Beahrs OH. American Association of Endocrine Surgeons. Presidential address: Lest we forget. Surgery 1987; 102(6):893-7.

5   Kegel AH. Domestic mechanical refrigeration in relation to public health. Ill. Med J 1930; 58:424-7.

6   Kegel AH. The health of the school child. Can J Med Surg 1931; 69:69-72.

7   Starr DP. Blood : an epic history of medicine and commerce. 1st ed. New York: Alfred A. Knopf, 1998.

8   Kegel AH. Stress incontinence of urine in women; physiologic treatment. J Int Coll Surg 1956; 25(4 Part 1):487-99.

9   Galli-Curci A. Let's talk about my operation. New York Journal Sept. 14, 1935.

10 Le Massena CE. Galli-Curci's life of song. New York: The Paebar Co., 1945.

11 Moosman DA, DeWeese MS. The external laryngeal nerve as related to thyroidectomy. Surg Gynecol Obstet 1968; 127(5):1011-6.

12 Roder CA. Operations on the upper pole of the thyroid. Arch Surg 1932; 24:426–439.

13 Arnold GE. Physiology and pathology of the cricothyroid muscle. Laryngoscope 1961; 71:687-753.

14 Abitbol J, Abitbol B. [The Voice and menopause: the twilight of the divas]. Contracept Fertil Sex 1998; 26(9):649-55.

15 Abitbol J, Abitbol P, Abitbol B. Sex hormones and the female voice. J Voice 1999; 13(3):424-46.

16 Stoicheff ML. Speaking fundamental frequency characteristics of nonsmoking female adults. J Speech Hear Res 1981; 24(3):437-41.

17 Oertli D, Udelsman R. Surgery of the thyroid and parathyroid glands. Berlin ; New York: Springer, 2007.

Table

Career span of famous coloratura sopranos. Data from individual biographies and operatic encyclopedias.

Name	Born	Debut	Retirement	Comments
Caballé, Montserrat	1933	1956	1989	More mature roles after 1989
Callas, Maria	1923	1950	1965	
Lind, Jenny	1820	1838	1849	recitals in USA in 1850s
Pagliughi, Lina	1907	1927	1957	
Pasta, Guiditta	1797	1826	1835	
Patti, Adele	1843	1851	1897	
Sills, Beverly	1929	1947	1980	
Sutherland, Joan	1926	1950	1989	more mature roles after 1970
Tacchianardi-Persani, Fanny	1812	1832	1852	
Tetrazzini, Luisa	1871	1907	1914	sang in recitals until 1934

10

Connell's Fundusectomy, Zollinger's "Dream" Ulcer Operation, and Two Princes of Serendip
The Discovery of the Zollinger – Ellison Syndrome

Stuart D. Wilson

Introduction

There is an old Persian fairy tale called *The Three Princes of Serendip*. The story tells of a great powerful king by the name of Giaffer and his efforts to prepare one of his three sons to succeed him as king of their Far East country which was called Serendip.

The three very intelligent sons were trained in the arts and sciences by the best tutors in the land. As part of their education and preparation to be leaders, the king sent them on a journey to another kingdom to acquire empirical experience. The fairy tale details the three princes' experiences and discoveries during their travels. They were constantly making discoveries, but purely by accident and of things they were not in quest of. The English author Horace Walpole (1717-1797) invented the word "serendipity" to describe the process of making discoveries when one is looking for something very different.

Richard Boyle describes, in detail, Horace Walpole's first use of the word serendipity and reminds us of other examples of accidental discovery while looking for something else; Christopher Columbus's discovery of America, Alexander Fleming's discovery of penicillin, and Alfred Bernhard Nobel's discovery of dynamite are cited.[1]

At the 1955 annual meeting of the American Surgical Association in Philadelphia, PA, Robert M. Zollinger and Edwin H. Ellison (Figure 1) presented two patients with primary peptic ulcers of the jejunum, gastric hypersecretion of gigantic proportions, and non-beta islet cell tumors. Their hypothesis was that "an ulcerogenic humoral factor of pancreatic islet origin" was responsible. Their report captured the imagination of and excited surgeons, clinicians, and physiologists worldwide.[2] Not unlike the Princes of Serendip, Zollinger

Figure 1: Robert M. Zollinger, Sr MD, (1904-1992) and Edwin H. Ellison MD (1918-1970) Ohio State University College of Medicine, 1954.

and Ellison had discovered, by accident and sagacity, something they were not in quest of; they were trying to understand the problem of primary jejunal ulcers and to find a better ulcer operation. The "ulcerogenic hormone" concept was their unexpected discovery.

This story of discovery is even more remarkable because the clinical experiences and coincidences that led to the discovery occurred during 12 short months at the Ohio State University Hospital in just two patients who had recurrent ulcers and numerous operations. This chapter focuses on the sequence of events in 1954 that led to the discovery of Zollinger- Ellison Syndrome. (Table 1)

Zollinger and Ellison's report begins with the "very trying clinical experiences" of two patients who presented with primary jejunal ulcers and stubborn ulcer recurrences despite all the traditionally accepted forms of medical and surgical treatments, short of total gastrectomy.[2]

January 19, 1954: Zollinger's jejunal ulcer patient JM admitted

Patient JM was a 19 year old girl first admitted to Zollinger's service on January 19, 1954, with abdominal pain, nausea, vomiting, and a possible bowel obstruction. She had a history of perforation of two separate jejunal ulcers which were closed primarily six months previously at another hospital. An

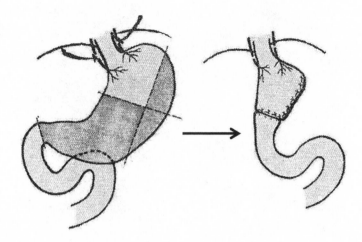

Figure 2: Illustration of Zollinger's "Dream Operation" for jejunal ulcers-: a subdia-phragmatic vagotomy, radical gastrectomy with fundusectomy and gastroduodenost-my. A 6cm x 8cm pouch remains. Connell's fundusectomy concept is incorporated.

initial upper barium radiologic study showed no ulcer but a 12 hour gastric aspiration produced a volume of 2800 ml with 300 mEq of acid. A repeat GI series suggested a duodenal ulcer and possibly a jejunal ulcer. Because of the gigantic amounts of hydrochloric acid being produced, Zollinger decreed "a heroic effort" would be made to control the marked gastric acid hypersecretion.[3] A more radical operation would be necessary to reduce gastric acid secretion.

January 28, 1954: Zollinger's "Dream Operation" documented in a movie

Zollinger's "Dream Operation" to control the marked gastric acid hypersecretion in patient JM was documented with a color motion picture. The plan was to submit the movie of this novel operation to the motion picture committee of the American College of Surgeons for the program at the next Clinical Congress of the American College of Surgeons. Zollinger had reasoned that a more conservative ulcer operation might not control the gigantic amounts of hydrochloric acid being produced. Accordingly, a subdiaphragmatic vagotomy, radical gastrectomy with fundusectomy, and end-to end gastroduodenostomy were performed (Figure 2). The residual gastric pouch measured only 6cm by 8cm. "The fundus of the stomach had been removed to cut down on the

Figure 3: F. Gregory Connell MD,
FACS, (1875-1968) Oshkosh,
Wisconsin

amount of gastric secretion, a procedure first recommended in 1931 by F. Gregory Connell (Figure 3) of Oshkosh, Wisconsin".[4] Interestingly, this comment and reference to Connell's fundusectomy research[4-10] was not mentioned in the historic 1955 presentation by Zollinger and Ellison, but was quoted 20 years later by Zollinger in his book "The Influences of Pancreatic Tumors on The Stomach".[3] Another interesting aspect to the story regarding Connell's research on fundusectomy was that when Zollinger was a young surgical resident in training at the Peter Bent Brigham Hospital in Boston he had done similar fundusectomy experiments in dogs. His paper entitled "Fundusectomy in the Treatment of Peptic Ulcer" was reported in Surgery, Gynecology and Obstetrics in 1935.[11] Zollinger's animal studies showed a gradual return of the free and total acid to preoperative levels when measured at one , three, and eight months after fundusectomy The paper concluded "the experimental findings agree with the clinical observation that the gastric acidity cannot be permanently lowered regardless of amount of acid bearing tissue removed" and "fundusectomy has little to offer as a means of permanently reducing gastric acidity and as a possible therapeutic measure in control of peptic ulcers".

However, Zollinger was backed into a corner in 1954, when JM developed recurrent ulcers and there was an urgency to choose a more aggressive ulcer operation short of a total gastrectomy for his young patient. Zollinger evoked Connell's fundusectomy principle for use in his "Dream Operation". His own previously reported animal studies and familiarity with the fundusectomy literature surely brought to mind that the greatest concentration of acid-producing parietal cells was located in the fundus of the stomach. Interestingly, although referencing Connell's work, Zollinger did not mention or cite his own 1935 publication regarding fundusectomy studies in dogs that suggested "fundusectomy has little to offer."[11]

February 27, 1954: Ellison's jejunal ulcer patient CP admitted

Patient CP, a 36 year-old woman, had previously been seen by Ellison in 1952. She had several jejunal sleeve resections for bleeding jejunal ulcers. Subsequent vagotomy and two gastric resections had failed to prevent recurrent ulcers. Twelve-hour overnight gastric acid measurements exceeded 100 meq. Subsequently, she was operated on several more times at another hospital. A transthoracic vagotomy was performed and later, more stomach was resected for a marginal ulcer. The postoperative course was complicated by a gastrocutaneous fistula and the patient was transferred to Ellison at the University Hospital on February 27, 1954, weighing only 72 pounds and very malnourished.

April 14, 1954: Ellison's patient CP has urgent total gastrectomy

Ellison performed an operation on April 7 to place a catheter and promote better drainage for a disrupted anastomosis and abscess cavity. Two weeks later, two successive massive gastrointestinal bleeds and 18 units of whole blood necessitated an urgent operation and a total gastrectomy was performed on April 14, 1954. The postoperative course was complicated by a pleural effusion and an anastomotic fistula draining large amounts of bile and jejunal contents, leading to her demise on June 7, 1954.

June 7, 1954: Autopsy patient CP, reported gross finding "normal pancreas"

The gross autopsy report was signed out "clinical diagnosis: recurrent jejunal ulcers, drainage of perigastric cavity post gastric resection with esphago-jejunostomy cutaneous fistula complication." It is significant that the initial autopsy report (gross only) stated that "the pancreas is of normal size, external and cut surfaces are lobulated, tan, firm and showed no significant changes".

Of note, no pancreatic tumors were seen by the pathologist.

June 14, 1954: Ellison's letter to CP's referring MD

One week after his patient's death, Ellison corresponded with the referring physician. His very detailed letter reviewed CP's seven previous operations for the jejunal and recurrent ulcers and the dramatic values measured in the gastric aspiration. The final histologic reports for the autopsy had not yet been completed. No mention was made of any pancreatic tumors in Ellsion's letter.

July 30, 1954: Zollinger's letter to Jenkins- "Dream Operation" failed, withdraw movie

Zollinger wrote Hilger P. Jenkins, the chairman of the motion picture committee of the American College of Surgeons indicating that the "Dream Operation" (Zollinger's terminology) for control of marked gastric hypersecretion had failed, and obviously the motion picture made at the time of fundusectomy, vagotomy, and gastroduodenostomy should not be shown.

Aug 5, 1954: Jenkins' quick reply: "Any sign of hyperinsulinism?, islet cell tumor?"

Jenkins responded within days to Zollinger's letter requesting withdrawal of their "Dream Operation" movie because patient JM had developed a recurrent ulcer shortly after the operation. Jenkins wrote back, "your case of primary jejunal ulcer intrigues me very much...This reminds me a bit of a case that I had of multiple gastric, duodenal and jejunal ulcerations and ultimate perforations in a lady that had an islet cell tumor of the pancreas. She was apparently carrying out a constant insulin test on her gastro-duodenal jejunal mechanism which was her undoing. Is there any possibility that your little girl has a touch of hyperinsulinism as an etiology for this interesting condition?"

Jenkins' reasoning as to the etiology of the jejunal ulcers is understandable. In the 1950s, patients with recurrent peptic ulcer after surgery frequently were tested for incomplete vagotomy with a "Hollander test." This test was based on the principle that gastric acid secretion can be stimulated via the intact vagal nerves by insulin-induced hypoglycemia. Jenkins logically proposed that a patient with an insulin-secreting islet cell tumor of the pancreas might

develop gastric acid hypersecretion. Zollinger later credited Jenkins as the first
to suggest they look for hyperinsulinism and an islet cell tumor. However,
subsequent studies in patient JM showed no evidence of an insulinoma or
insulin-induced increased gastric acid secretion.

October 8, 1954: Clinical Pathology Conference (CPC). Autopsy in patient CP now reports islet cell tumor

Less than eight weeks after Jenkins' letter suggested an islet cell tumor, Ellison's patient CP, who had died postoperatively, was the case selected for
the weekly University Hospital CPC conference. The traditional CPC teaching conference consisted of a presentation of a case history without revealing the final diagnosis. A faculty clinician was assigned to discuss the case,
demonstrating the reasoning process that would lead to the correct diagnosis.
Students or residents in the audience might then also vote on a possible diagnosis. The pathologist would then present the correct anatomic diagnosis
based on tissue removed during an operation or autopsy. A discussion usually
followed in order to critique the diagnosis offered by the invited clinician. For
this particular CPC, Ellison was selected to discuss the case history of his
jejunal ulcer patient, CP, who had succumbed after numerous heroic attempts
(seven operations) to control the ulcer diathesis and continued bleeding. It is
interesting to note that following Ellison's discussion of jejunal ulcers and the
dramatic gastric acid hypersecretion, the pathologist revealed for the first time
that several islet cell tumors were identified in the pancreas. There is no record
to know if the pathologist asked Ellison if he had noticed any pancreatic tumors in the body and tail of the pancreas during the previous operations. It is
understandable why Ellison did not appreciate the small islet cell tumors since
the pathologist also did not appreciate these lesions at the time of autopsy
when he removed the pancreas. Indeed, the tumors were not found until the
pancreas was sectioned for histologic examination.

Remember that the gross autopsy report on June 7, 1954 after CP's death
was signed out "Clinical diagnosis: recurrent jejunal ulcers, drainage of perigastric cavity post gastric resection with esophagus-jejunostomy, cutaneous
fistula complication" and the record stated "the pancreas is of normal size,
external and cut surfaces are lobulated, tan, firm and showed no significant
changes." Ellison's letter to CP's referring physician was dated one week after
the autopsy and he makes no mention of any pancreatic tumors. The timing
of this new finding of pancreatic islet cell tumors in Ellison's patient, which

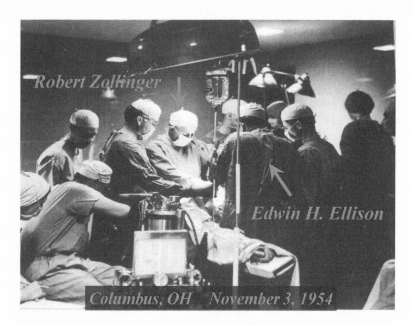

Figure 4: Zollinger and Ellison performing a total gastrectomy November 3, 1954 on patient JM after "Dream Operation" failed. Pancreatic tumors searched for and found.

was reported by the pathologist at the CPC, is remarkable. Jenkins' letter about his young patient with jejunal ulcers suggesting Zollinger look for an islet cell tumor in patient JM as a reason for failure of his "Dream Operation" had arrived just weeks before the CPC. The finding of the islet cell tumors in patient CP, apparently reported for the first time at the CPC surely must have been a "Eureka moment" for Ellison. Ellison had developed a special interest in the uncommon problem of jejunal ulcers since his patient as well as that of Zollinger were on the surgical ward at the same time in February 1954. Both had had jejunal perforations and marked gastric acid hypersecretion. Jenkins' letter questioning the possibility of an associated islet cell tumor in patients with jejunal ulcer arrived, and now the serendipitous finding of pancreatic tumors in patient CP.

November 3, 1954: Zollinger Reoperates JM -"Dream Operation" had failed. Total gastrectomy, searches and finds pancreatic tumors

Zollinger reoperated on patient JM November 3, 1954, planning a total gastrectomy as a last resort to control the persistent ulcer diathesis (Figure 4).

The operation was performed less than four weeks after the CPC for Ellison's patient and the "Eureka moment" when the pathologist reported the finding of unsuspected islet-cell tumors in the tail of the pancreas. Remember too that Jenkins' letter relating the story of his patient with jejunal ulcers in which he suggested that Zollinger look for an islet cell tumor as an etiology for the ulcer diathesis had arrived shortly before the CPC.

Because of Jenkins' suggestion about a hyperinsulism theory, Zollinger's patient JM was fasted for 48 hours looking for evidence of an insulinoma, but hypoglycemia was not documented and blood sugars remained above 80 mg. Thus, there was of no clinical evidence of an insulinoma.

At operation, a large posterior penetrating esophageal ulcer and a second posterior penetrating 4 x 3 cm ulcer at the gastroduodenal anastomosis were found. A total gastrectomy was performed and the pancreas was inspected for tumors. With great expectation and excitement, the search proved rewarding. Two small tumor nodules, initially thought to be lymph nodes, were removed from the anterior aspect of the tail of the pancreas. It was concluded that one of the tumor nodules was an islet cell tumor in the pancreas and the second lesion was a lymph node with metastatic islet cell tumor, non-beta cell type.

November 14, 1954: Surgical Biology Club I: Ellison's presentation "The Problem of Jejunal Ulcers", "Ulcerogenic factor of pancreatic origin" concept suggested for the first time

The presentation at the Surgical Biology Club I, "The Problem of Jejunal Ulcers" by Ellison was just days after Zollinger's operation finding islet cell tumors in patient JM. The audience was a small group of leaders in American surgery (Figure 5) and they heard the case histories of the two patients with recurrent jejunal ulcers and "gastric acid hypersecretion of gigantic proportions despite adequate conventional medical or surgical therapy". This was the first presentation to any group postulating "an ulcerogenic factor of pancreatic islet origin" and suggesting "a new clinical entity consisting of hypersecretion, hyperacidity, and atypical peptic ulceration associated with non–insulin-producing islet cell tumors of the pancreas".

December 1, 1954: Abstract submitted to the American Surgical Association

An abstract for the paper entitled "Primary Peptic Ulceration of the Jeju-

Figure 5: Surgical biology Club I, Hackney's Seafood Restaurant, Atlantic City November 14, 1954. Edward Ellison, Francis Moore, Henry Randall, William Holden, and Curtis Artz are identified. They provided Ellison support for the ulcerogenic tumor concept.

num Associated with Islet Cell Tumors of the Pancreas" was submitted to the American Surgical Association program committee for consideration. The abstract was accepted for presentation before the annual meeting in Philadelphia, PA April 29, 1955.

American Surgical Association Presentation

The now historic paper entitled "Primary Peptic Ulcerations of the Jejunum Associated with Islet Cell Tumors" was presented at the annual meeting of the American Surgical Association in Philadelphia on April 29, 1955. Ellison presented the paper, although he was listed as the second author. Zollinger answered the questions during the discussion period that followed.

The five discussants for the paper were Hilger P. Jenkins, Chicago, IL; Allen O. Whipple, Princeton, NJ; Lester Dragstedt, Chicago, IL; Edgar J. Poth, Galveston, TX; and Carl A. Moyer, St. Louis, MO. All five were prominent leaders in American Surgery. They were invited to discuss the paper months before the meeting by Ellison. The choice of these particular surgeons illustrates the detailed planning for their presentation by the authors in order to

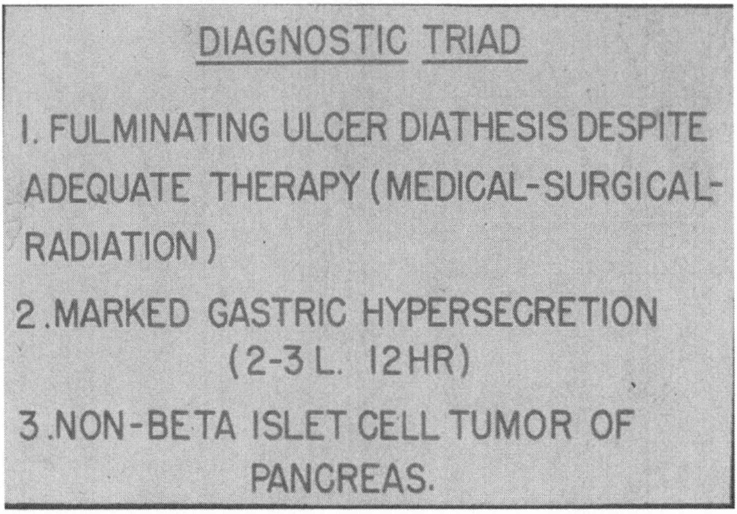

Figure 6: The glass lantern slide that Ellison projected at the 1955 American Surgical Association meeting. The "diagnostic triad" for a new clinical syndrome.

promote the proposed idea that the syndrome was a new clinical entity.

Jenkins discussed his own jejunal ulcer patient with allegedly the same syndrome and reported that an islet cell tumor was found at autopsy. Zollinger credited him for being the first to suggest they look for a pancreatic tumor.

Whipple's invitation was largely because of his classic description of the triad of symptoms found in patients with insulin-producing pancreatic tumors, the so-called "Whipple triad". It is notable that Ellison had a slide in his presentation labeled "Diagnostic Triad" for the ulcerogenic tumors of the pancreas syndrome (Figure 6).

Dragstedt's comments were to prove prophetic. After describing one of his own cases of fatal ulcer hemorrhage and finding an islet cell pancreatic tumor at autopsy, he then discussed his studies of ulcer disease due to hypersecretion of gastric juice of humoral origin resulting from hyperfunction of the gastric antrum. He commented "it is quite possible that the study of these pancreatic cells may provide clues to the type of cells in the gastric antrum that produce gastrin."

Poth was well known for his ongoing investigations about the relationship of the pancreas and peptic ulcers. He discussed one of his patients with recurrent ulcer disease and the effects of glucagon administration. Ellison had been corresponding with Poth about the possibility of a role for glucagon in the jejunal ulcer syndrome; however, no definite conclusions were made from his research.

Moyer was the final discussant and described two St. Louis patients with the proposed diagnostic triad concluding "the association of duodenal ulcer and jejunal ulceration after gastric resection with islet cell carcinoma seems to be a clinical entity...." Moyer acknowledged that Ellison had prepared the slides with the sketches of his patients' operations shown during his discussion.

The Eponym: The Zollinger –Ellison Syndrome

Ben Eiseman, a young guest surgeon from Denver, Colorado was in the audience during the American Surgical Association presentation, and he recognized that the clinical triad proposed by Drs. Zollinger and Ellison was present in one of his patients. He was able to diagnose this new entity preoperatively in a patient at the Denver VA hospital. He returned to Denver, and the following week performed a total gastrectomy and distal pancreatectomy in a patient with an islet cell carcinoma. His patient had undergone five previous operations attempting to cure recurrent ulcer disease. Drs. B. Eiseman and R.M. Maynard published their case in Gastroenterology (submitted November 19, 1955, published September 1956). In their published report, they were the first to suggest "for the sake of simplicity, we propose this clinical entity be called the Zollinger-Ellison Syndrome."[12]

In less than a year, Ellison had collected 24 patients with ulcerogenic tumors for a second report further defining clinical histories and pathology of this new syndrome. Ellison presented the report at the annual meeting of the Society of University Surgeons, Indianapolis, Indiana, in February 1956. The paper was entitled "The Ulcerogenic Tumor of the Pancreas" and Ellison was the single author. Eiseman was once again in the audience, but this time, he was a discussant reporting information about his patient with an ulcerogenic tumor, and also, that his preliminary dog studies "suggested that glucagon probably is not an ulcerogenic factor in this interesting disease." Eiseman again suggested "that this clinical entity be called the Zollinger-Ellison Syndrome."[12] This eponym, the Zollinger–Ellison syndrome or Z-E syndrome caught on, and has been used in the heading of thousands of case reports since. PubMed now lists more than 3400 citations. Zollinger did not use this eponym in his subsequent reports, preferring the term "ulcerogenic tumor of the pancreas."

Following the discovery that gastrin was the responsible hormone and was the gastric secretagogue elaborated by the islet cell tumors, these ulcerogenic tumors are now appropriately called "gastrinomas."[13, 14]

Epilogue

Fifty years have passed since Zollinger and Ellison presented their classic paper. During the next several decades (1955-1975), subsequent reports of patients with ulcerogenic tumors, the discovery of a potent gastric secretagogue in this tumor and in the blood of Z-E patients, the identification of gastrin as the ulcerogenic hormone and development of radioimmunoassay methods to measure physiologic serum gastrin concentrations, all make fascinating reading. The history of these first discoveries has been recorded by Zollinger along with detailed clinical histories of the first two patients, their numerous failed ulcer operations and the comment by Zollinger relating that "at frequent intervals during 1954, Ellison and I sought consultation as well as consolation with each other."[3]

During the next three decades (1975-2005) a better understanding of the pathophysiology of this disease, the advent of effective pharmacologic gastric acid inhibitors, improved imaging modalities to localize tumors, and long-term outcome results of different operative and medical strategies have been responsible for the evolving approaches to the diagnosis and treatment of this disease.

The syndrome continues to captivate many physician-scientists. Some have made the disease entity a focus of their research and clinical practice, and they might appropriately be called "Z-E Watchers". Their contributions continue to advance our understanding. A number of reviews outline the historical evolution of the Z-E syndrome and cite the numerous contributors.[3, 13-17] A recent monograph "The Zollinger- Ellison Syndrome: A Comprehensive Review of Historical, Scientific, and Clinical Considerations" is recommended as a most complete reference.[16] Of historical note, the author, E. Christopher Ellison, MD is the son of the late Edwin H. Ellison, MD and is currently the Robert M. Zollinger Professor and Chair, Department of Surgery, Ohio State University Medical Center, Columbus, Ohio- the original site of this discovery story. His continuing contributions to this syndrome have been significant.

Zollinger and Ellison's concept of an "ulcerogenic hormone" of pancreatic islet cell origin and related peptic ulcer disease stimulated an explosion of new studies and initiated a brotherhood of "Z-E watchers," investigating the pathophysiology, diagnostic avenues and best outcome treatment for this new disease entity, the Z-E syndrome. The unique coincidence of Connell's and Zollinger's early dog experiments regarding fundusectomy operations to reduce gastric acid secretions, followed years later by the clinical sagas of Zollinger and Ellison's two patients with primary jejunal ulcers, their failed ulcer operations leading to total gastrectomy, and the unexpected finding of associated pancreatic islet cell tumors set the stage for discovery. Zollinger and Ellison might well be called our Princes of Serendip of a more modern era.

TABLE 1

1954 JEJUNAL ULCER EXPERIENCES
CHRONOLOGY OF EVENTS FOR THE PRINCES OF SERENDIP

Jan. 19	Zollinger's jejunal ulcer patient JM admitted.
Jan. 28	Zollinger's "Dream Operation" movie-patient JM.
Feb. 27	Ellison's jejunal ulcer patient CP admitted.
Apr. 14	Ellison's patient CP has urgent total gastrectomy. 7 previous operations had failed, continued bleeding.
Jun. 7	Autopsy patient CP, reported gross finding "normal pancreas".
Jun. 14	Ellison's letter to CP's referring MD, no mention of any pancreatic tumors.
Jul. 30	Zollinger's letter to Jenkins Our "Dream Operation" for jejunal ulcers failed, withdraw movie.
Aug. 5	Jenkin's quick reply to Zollinger. "Any sign of hyperinsulinism" he suggests possible islet cell tumor.
Oct. 8	Clinical Pathology Conference: Ellison selected to discuss patient CP. Pathologist's final autopsy now reports -islet cell tumors of pancreas.
Nov. 3	Zollinger reoperates JM- the "Dream Operation" had failed, Total Gastrectomy, searches and finds pancreatic tumors.
Nov. 14	Surgical Biology Club I: Ellison's presentation "The Problem of Jejunal Ulcers". "Ulcerogenic factor of pancreatic origin" concept suggested for the first time.
Dec. 1	Abstract submitted for American Surgical Association annual meeting. "Primary Peptic Ulcerations of the Jejunum Associated with Islet Cell Tumors of the Pancreas".

References

1   Boyle R. The Three Princes of Serendip 2000. Available at: http://
    livingheritage.org/three_princes.htm#. Accessed 7-03, 2012.
2   Zollinger RM, Ellison EH. Primary peptic ulcerations of the jejunum
    associated with islet cell tumors of the pancreas. *Ann Surg* 1955;
    142(4):709-23; discussion, 724-8.
3   Zollinger RM, Coleman DW. The Influences of Pancreatic Tumors on
    the Stomach. Springfield: Charles C Thomas, 1974.
4   Connell FG. Fundusectomy. *Surgery, Gynecology and Obstetrics* 1931;
    53(December):750-752.
5   Connell FG. Fundusectomy. *Surgery, Gynecology and Obstetrics* 1929;
    Xlil(No.5):696.
6   Connell FG. Resection of the Fundus of the Stomach for Peptic Ulcer.
    Ann Surg 1932; 96(2):200-3.
7   Connell FG. Partial Gastric Fundusectomy inTreatment of Peptic Ulcer
    Surgery, Gynecology and Obstetrics 1934; 59:786-788.
8   Connell FG. Partial Gastrectomy in Treatment of Peptic Ulcer. *Surgery*
    1938; 3:696-702.
9   Connell FG. Partial Fundusectomy (Proximal Gastrectomy) : Review of
    24 Cases. *Ann Surg* 1943; 118(6):1000-4.
10 Connell FG. Trends in treatment of duodenal ulcer; partial
    fundusectomy; results in 25 cases. *Rev Gastroenterol* 1948; 15(3):247-9.
11 Seely H, Zollinger RM. Fundesectomy in theTreatment of Peptic Ulcer.
    Surgery, Gynecology and Obstetrics 1935:155-161.
12 Eiseman B, Maynard RM. A noninsulin producing islet cell adenoma
    associated with progressive peptic ulceration (the Zollinger-Ellison
    syndrome). *Gastroenterology* 1956; 31(3):296-304.
13 Wilson S. Gastrinoma (Zollinger-Ellison Syndrome, Ulcerogenic Tumor
    of the pancreas). In: Howard J, Jordan Jr. GL, Reber HA, ed. Surgical
    diseases of the pancreas. Philadelphia: Lea & Febiger; 1987:pp. 829-847.
14 Wilson S. Gastrinoma. In: Clark OH, Siperstein AE, Duh QY, eds.
    Textbook of Endocrine Surgery. Orlando, FL: W. B. Saunders Co;
    1997:pp. 607-618.
15 Landor J. The Zollinger-Ellison Syndrome. *In* LLoyd M. Nyhus M, ed.
    Problems in General Surgery, Vol. 7. Philadelphia: J. B. Lippincott 1990.

16 Ellison EC, Johnson JA. The Zollinger-Ellison syndrome: a comprehensive review of historical, scientific, and clinical considerations. *Curr Probl Surg* 2009; 46(1):13-106.
17 Ellison EH, Wilson SD. The Zollinger-Ellison Syndrome:Re-appraisal and evaluation of 260 registered cases. *Ann Surg* 1964; 160:512-30.

11

The Identification of Primary Aldosteronism
Jerome Conn and His First Patient

Norman W. Thompson and Roger J. Grekin

In the annals of medical discovery, rarely has an individual been better prepared when the opportunity arose than was Jerome Conn when referred a patient in April, 1954 with what would later become known as Conn's Syndrome. As he stated years later, "that patient could not have been presented to anyone more conscious of the possibility of aldosteronism in man than I was at that moment."[1, 2]

Jerome Conn's Academic Background

In April 1954, when this story began, Jerome Conn was Professor of Medicine, Chief of the Division of Endocrinology and Metabolism and the Director of the Metabolic Research Unit at the University of Michigan. Jerry, as he was known, was born in 1907 in Manhattan, where his parents were shopkeepers. He completed his undergraduate work in three years at Rutgers University, after which he was accepted to the University of Michigan Medical School in 1928. After graduating with honors in 1932, he remained at the University Hospital for a surgical internship under Frederick A. Coller. He then changed his interests from surgery to more basic clinical investigations and completed a two-year residency in internal medicine, followed by a fellowship in the Division of Clinical Investigation. There, he was intrigued by exciting research in obesity, energy metabolism and diabetes under the direction of Louis H. Newburgh. In medical school Jerry met Betty Stern, a classmate who shared his interest in clinical research. They married his first year of medical school and she then joined him in the same division as a co-investigator in research projects involving non-insulin-dependent diabetes. After three years, he was appointed Assistant Professor (1938) of Medicine. Five years later he was promoted to the position of Chief of the Division of Endocrinology and Metabolism when Newburgh left Michigan for the United States Naval

Investigation Laboratory in Washington, D.C.

Conn held this position for the next 30 years. He was also the Director of the Hospital's Metabolism Research Unit that was fully supported by external grants for clinical studies lasting for several weeks or longer.[3] There was also a sophisticated biochemical laboratory and full-time research dietician allowing for state-of-the-art metabolic investigations.

Acclimation to Warm Climate Studies

Shortly after the United States entered World War II, the acclimation of military personnel to warm climates became of great practical importance to the armed forces. In 1943, Jerry's metabolic research unit was one of several selected for these important studies and his response was immediate. A tropical climate room maintained at 90° F with a humidity of 80-90% was constructed and a group of conscientious objectors were recruited as subjects to be studied in the hot room for periods up to 90 days. The process of adaptation was found to require ten to twenty days to complete. Changes in cardiovascular function and in the function of the sweat glands are the factors that are critical. With acclimation, there is maintenance of an increased plasma volume. In moist heat, a large volume of sweat evaporates, but does not decrease body heat. There is also a marked diminution of sodium chloride in sweat. From an initial sweat loss of 5 to 7 liters per day containing 10 to 14 grams of sodium chloride, there is a decrease to about 1 gram per day by the tenth day. The kidneys also conserve NaCl acutely but urinary sodium eventually rises to match sodium intake after ten days when a new steady state is achieved. In the early phase, the mechanism was believed due to an acute increase in adrenal cortical activity by a deoxycorticosterone-like hormone. In man, 7 mg/day of deoxycorticosterone produced the same changes in sweat and urine in unacclimated subjects.[1] In retrospect, the salt-saving steroid being studied was aldosterone. Studies showed that the mechanism does not require an increased ACTH production. It seemed clear that the adrenal cortical elaboration of increased amounts of a deoxycorticosterone-like material constituted the major factor in man's ability to cope with a hot environment by reducing NaCl losses from the skin by as much as 95%.[2,4-9] It should be of interest, particularly to surgeons, that during this time, Conn also was collaborating with Frederick Coller's laboratory research group studying post- operative fluid and electrolyte requirements. They found that salt retention lasted for about eight days after the stress of a major abdominal operation and was caused by the selective increased secretion of an adrenal mineralocorticoid.[10] This led to a more rational approach for fluid and electrolyte replacement in post-operative patients and was considered a major contribution to the field of surgery.

Conn later stated that his acclimation research was more than casually related to primary aldosteronism but actually intimately related in presenting a fascinating combination of circumstances.[1, 2] It should be noted that all of these studies were done before the detection of electrocortin (aldosterone) by Grundy, Simpson and Tait in 1952.[11] In rapid succession aldosterone was isolated, chemically identified and synthesized by 1954.[12-14] Because of Conn's studies in the hormonal regulation of salt excretion in acclimation, he was indeed exceptionally well-prepared when confronted with the first patient found to have an excess secretion of aldosterone.

The First Patient with Primary Aldosteronism

MW, a 34 year-old woman, was initially seen at the University of Michigan outpatient clinic in April, 1954 and admitted to the internal medicine service with a diagnosis of "periodic paralysis," probably due to hypoparathyroidism." She was subsequently referred to Endocrinology and Metabolism, where she was seen in consultation by Stefan Fajans, a young faculty member. After an initial presentation by the resident, Fajans recorded his impressions in his own handwriting: *Hypokalemic Alkalosis, etiology to be determined. Alkalosis may be responsible for tetany. Usual etiological factors for potassium depletion are not present. Question of either excess adrenal cortical function in respect to electrolyte control only, or specific renal tubular defect with potassium loss to be considered. Advise: patient to be admitted to Research Lab for study.*[15] She had complained of muscle weakness, intermittent muscle spasms and even paralysis during the previous 7 years. In 1947 she had been paralyzed from the hips down for 2 full days and had had shorter intermittent periods of tetany since. Polyuria and nocturia were also significant symptoms throughout the 7 years. For the previous 4 years, she was persistently hypertensive in the 180-190/100-105 mm Hg range. Initial studies showed severe hypokalemic alkalosis, hypernatremia and impaired renal tubular absorption of water. Her initial serum potassium levels ranged from 1.5 to 2.4 meq/L, sodium 146-152 meq/L and the urinary pH was 7.62. The blood pressure was 170/110 mm Hg. She had no edema. The Chvostek's and Trousseau's signs were positive. She was then presented to Jerry Conn who fully concurred with Fajans' diagnosis of suspected mineralocorticoid hypersecretion. He suggested immediately that the metabolic abnormality of severe hypokalemia, mild hypernatremia and alkalosis could result from excessive secretion of a salt-active adrenal steroid. He emphasized the singularity of the mineralocorticoid excess because of the absence of any findings suggesting an increase in either cortisol or androgen production.[15, 16] Fajans was to have a long and distinguished career at Michigan, but with a primary focus on diabetes research. He succeeded Conn as Chief of the Division of Endocrinology and Metabolism in 1973[3] and submitted his last scientific paper for publication in October 2011 at age 92 years.[15]

Conn had MW admitted to the Metabolic Research Unit on April 27, 1954, where the first crucial studies consisted of a series of thermal sweat salt levels. As Conn suspected, these showed very low sodium levels. One study lasted for 35 days. Sweat and salivary sodium were greatly depressed whereas sweat potassium levels were very high. Attempts to raise serum potassium levels with a high intake of potassium resulted in only transient improvement. Levels of 17-hydroxy and ketosteroids were consistently normal, indicating normal cortisol and adrenal androgen secretion. A young associate in the laboratory, Dr. David Streeten, found that the patient's urine contained an excess of mineralcorticoid, as demonstrated in a bioassay he developed using adrenalectomized rats. When compared with 7 normal subjects, MW's urine had greatly elevated levels. These and many other studies convinced Conn that the patient was suffering from mineralocorticoid excess and that an adrenalectomy was indicated.[9] Conn proposed a major surgical intervention with the establishment of iatrogenic permanent adrenal insufficiency as a cure for a previously unrecognized disease. This procedure had never been previously done under these circumstances. Conn did not hesitate to acknowledge and accept full moral responsibility for the treatment program he was advocating. According to Fajans, he requested any dissenting opinions from any members of the Endocrine Division. There was a short discussion about the seriousness of bilateral adrenalectomy and steroid replacement therapy but there were no dissenting opinions about his proposed therapeutic plan.[15, 16]

The First Publications Establishing Primary Aldosteronism as a New Syndrome

The way in which this new syndrome was made known to the medical profession is particularly fascinating. Jerry was the president of the Central Society for Clinical Research and his presidential address was scheduled for October 29, 1954. He used this occasion to present for the first time his extensive clini-

cal investigations of this new syndrome which he entitled Primary Aldosteronism.[9] He concluded by stating that he believed that the patient should be
explored and most likely would require a bilateral adrenalectomy and life-long
steroid replacement. He predicted that this should abolish the entire metabolic
abnormality. He noted that the patient was currently in the hospital in preparation for this procedure. This presentation was published just three months
later in the *Journal of Laboratory and Clinical Medicine* with two addenda. The first
noted a paper from England entitled "Potassium-Losing Nephritis Presenting
as a Case of Periodic Paralysis" (Brit. Med.J. 2:1, 067, 1954) in which bioassays of electrocortin excretion gave abnormally high values. The rest of the
data also indicated that the patient studied had the same abnormalities that
Conn had described as being associated with Primary Aldosteronism.[17] The
short second addendum noted that MW had been operated on Dec. 10, 1954
and a 4 cm right adrenal cortical tumor was removed, leaving the left adrenal gland in situ. It is quite remarkable that this preliminary report of MW's
operation, even as an addendum, was published within less than a month of
her procedure. This may have been facilitated by the chief editor of the *Journal of Clinical and Laboratory Investigation* who was William Robinson, and was
coincidentally also Chairman of the Department of Internal Medicine at the
University of Michigan. Just three months later, in April 1955, Conn's more
detailed follow up report was published in the same journal.[18]

The Surgical Exploration and its Surprising Findings

The surgeon who performed this historic operation was William C. Baum,
a junior faculty member in the section of Urology. Why Reed Nesbitt, who
was the Chief of Urology and would subsequently do all of the other early
operations for aldosteronism at Michigan, was not directly involved is unclear.
Apparently he did make an appearance in the operating room when the adenoma had been discovered, but did not otherwise participate.[15] Baum, who
was one of the author's (NWT) instructors in Urology the following year,
left the University for private practice shortly thereafter. His name appeared
only in Conn's followup report.[18] Nesbitt authored all of the papers on the
surgical treatment of aldosteronism from Ann Arbor and was considered an
expert in this field. He was also internationally renowned for his pioneering
work on transurethral resection of the prostate and later served as president
of the American College of Surgeons.[19] Although Baum performed MW's
operation, it was personally directed by Jerry Conn who was present in the

operating room throughout. The operation was done through posterior, bi-lateral subcostal incisions, excising the 12th ribs. While the initial intent was to perform a bilateral adrenalectomy, once a 4 cm cortical tumor was found in the right adrenal gland, that plan was abandoned in favor of a right adrenalectomy and biopsy of the left adrenal. Conn and his colleagues were elated with finding an unsuspected cortical adenoma. At the time of this procedure, imaging techniques that could have identified this lesion preoperatively had not been developed. A 4 cm aldosteronoma could easily be detected by a number of radiologic, radio-isotopic, ultrasonic or selective venous assay methods in subsequent years. It should be emphasized that a 4 cm benign aldosteronoma is at the extreme end of size range for this disease. In an extensive operating experience with primary aldosteronism, NWT never encountered a benign aldosteronoma larger than 3 cm in diameter. During Conn's era, 3 large muscle biopsies were obtained in order to measure salt content before administering any intravenous hydrocortisone. Biopsies of both kidneys were also obtained before completing the operation.[18]

The round, encapsulated cortical adenoma was bright canary yellow color and was 14.8 grams in weight. Microscopy showed a benign cortical adenoma and atrophy of the zona glomerulosa of both adrenal glands. Bioassays of tumor extracts showed a marked increase in mineralocorticoid activity. These levels were 75 to 150 times greater per gram of tissue than that in beef adrenal. The wedge biopsy of the left adrenal showed atrophy limited to the zona glomerulosa that was one half as wide as normal. Previous studies had shown no depression of conventional endogenous ACTH production, again suggesting an aldosterone-secreting mechanism independent of pituitary regulation. Muscle biopsies showed a marked increase in intracellular sodium and a marked decrease in intracellular potassium. The renal biopsies showed severe arteriosclerosis and diffuse vacuolar changes in the renal tubular epithelium consistent with hydropic degeneration. Conn considered that this was due to chronic hypokalemia.[16]

Initial Followup and Subsequent Results

The patient's postoperative course showed a remarkable reversal of all metabolic abnormalities. Initially there was a large diuresis of sodium and sharp retention of potassium. 90% of these salt corrections occurred during the first 10 days. By the sixth postoperative day, the serum sodium and potassium levels returned to normal and remained so. The blood pH fell gradually from 7.55 to

7.42 by the 12th postoperative day. Her polyuria and polydipsia that had been so prominent disappeared on the 5th postoperative day. Her blood pressure, which averaged 170/100, began to fall gradually on the 12th day and by day 18 was 120/70 and remained there. Bioassays for aldosterone were normal. The patient became totally asymptomatic and had normal muscle strength.

Conn concluded that the syndrome of primary aldosteronism was already well established by his first case and the fact that within 6 weeks of his initial report, four additional cases had been recognized by others and in each a cortical adenoma had been found. These patients were all cured. In the case of Wyngaareden referred to in Conn's original report, no adenoma was found at autopsy.[20] He suggested that subtotal adrenalectomy be considered in cases in which no tumor was found. He considered that serum potassium and bicarbonate levels should be done as screening in all hypertensive patients because this new entity was probably more common that suspected. It could be easily detected and subsequently treated.[18] It is noteworthy that Conn's patient with periodic paralysis and a 4 cm adenoma represented the extreme end of the manifestations of this endocrine disease when discovered. This is similar to a number of other endocrine tumor syndromes when first discovered such as hyperparathyroidism (Mandl's patient with osteitis fibrosa cystica and a 2 cm. parathyroid tumor, probably malignant), insulinoma (Mayo's patient with intractable hypoglycemia and liver metastases) gastrinomas (Zollinger and Ellison's two patients with intractable peptic ulcer disease refractory to multiple surgical procedures short of total gastrectomy) and glucagonoma (Polk's patient with necrolytic migratory erythema and liver metastases from an alpha-cell tumor). Fortunately Conn's patient had a benign adenoma and not a rare aldosterone-secreting cortical carcinoma.

In the following decade, Conn's clinic became a referral center for patients with aldosteronism. Many subsequent publications expanded on the clinical description, laboratory and radiological diagnosis and surgical treatment of this and related conditions. When the work of others established that the secretion of renin-angiotensin was related to aldosterone, Conn's group published a series of studies of this system in hypertension and other conditions. He also studied the renin-angiotensin system in secondary aldosteronism that could be difficult to distinguish from primary aldosteronism. He was the first to report that a suppressed serum renin activity was an important distinguishing laboratory finding in primary aldosteronism due to an adrenal tumor.[21-25]

Although nuances of the disease and related disorders such as idiopathic aldosteronism were yet to be determined, Conn almost single handedly defined the syndrome of primary aldosteronism from the meticulous clinical and laboratory studies in a single patient. His experience, knowledge, insights and intuition led him to correctly focus on the adrenal glands as the source of his patient's disease. His convictions led him to courageously recommend an exploration of the adrenal glands and the discovery of an aldosterone-secreting adenoma, the excision of which cured the patient. With his publications, the medical world was made aware of an entity that had previously been totally overlooked.[9, 18]

Not only did he prove its existence, he showed how easily it could be diagnosed and treated. Many others would subsequently benefit from the "discovery" of the first patient to be diagnosed with primary aldosteronism, justifiably called Conn's syndrome by the medical world.

References

1   Conn, J.: Some Cllinical and ClimatologicalAspects of Aldosteronism in Man(The Gordon Wilson Lecture): Trans Am Clin Climatol Assoc. 1963; 74:61-91

2   Conn,J.W.: The Evolution of Primary Aldosteronism. 1954-1967. Harvey Lectures 1968; 62:257-291

3   Daughady, William H.: Jerome W. Conn 1907-1994, A Biographical Memoir 1997 National Acadamies Press, Washington. D.C.: pp.3-11

4   Conn, J.W.: Evolution of primary aldosteronism as a highly specific clinical entity: J. Am. Med. Assoc. 1960; April 9:172:1650-1655

5   Conn, J.W.: Electrolyte Composition of Sweat: Clinical implications as an index of adrenal cortical function: Arch.Int.Med. 1949; 83:416-428

6   Conn, J.W., L.H. Louis: Production of "Salt-active" corticoids as reflected in the concentrations of sodium and chloride of thermal sweat J.Clin Endocr.1950;10:12-23

7   Conn, J.W.,and Johnston, MW: The function of the sweat glands in the Economy of NaCl under conditions of Hard Work in a Tropical Climate: J.E. in. Envest. 1944;23:242-249

8   Conn, J.W., Johnston, MW, and Louis, L.H.: Acclimation to Humid Heat: A Function of Adrenal Cortical Activity. J. Clin. Invest. 1946;25:912

9   Conn, J.W. and Fajans, S.S.: The Pituitary-Adrenal System Annual Review of Physiology. 1952;14:453-480

10 Conn, J.W.: Presidential address. I Painting Background II Primary

Aldosteronism, a new clinical syndrome J. Lab & Clin. Med.45:1955; 45: 3-17

11 Johnson, H.T., Conn, J.W., Iob,V., and Coller, F.A. Postoperative salt retention and its relation to Increased Adrenal Cortical Function. Ann. Surg. 1950; 132:374

12 Grundy, H.M., Simpson, S.A. and Tait, J.F.: Isolation of a highly active Mineral corticoid from Beef Adrenal Extract. Nature 1952; 169:795

13 Simpson, S.A., Tait, J.F., and Bush I.E.: Secretion of a Salt-Retaining Hormone by the Mammalian Adrenal cortex Lancet 1952; 2:226

14 Simpson, S.A., Tait, J.F., Wettstein, A., Neber, R., Euw, J.Von, Schindler, O. and Reichstein, T.: Konstitution des Aldosterons, des neuen Mineralo corticoids. Experientia 1954; 10:132

15 Wettstein, A., et al.: Synthesis of Aldosterone. A paper presented at: Proc. International Congress of Pure and Applied Chemistry. Zurich, Switzerland (July 22) 1955

16 Fajans, S. Personal Communication (Oct., 2011)

17 Gittler, R.D. and Fajans, S.S. Prismatic Case: Primary Aldosteronism (Conn's Syndrome) J Clin Endocrand Metab 1995; 80: 3438-3441

18 Evans, B.M., and Milne, M.O.: Potassium Losing Nephritis Presenting as a case of Periodic Paralysis: Brit. Med. J. 1954; 2: 1067

19 Conn, J.W.: Primary Aldosteronism (Follow-up Report) J. Lab. Clini. Med. 1955; 45: 661-664

20 Conn, J.W., Knoff, RF, Nesbit, R.M.: Clinical Characteristics of Primary Aldosteronism from an analysis of 145 cases: Am. J. Surg. 1964; Jan: 107: 159-172

21 Wyngaarden, J.B, Keitel, J.G. and Isselbacher, J.: Potassium Depletion and Alkalosis. Their Association with Hypertension and Renal Insufficiences. N Engl J. Med. 1954; 250,597

22 Mandl, F.: The rapeuticher Versuch bei Osteitis fibrosa Generalisata mittels Esterpation eines Epithelkorperchentumors: Wien Lkin Wochensehr 1925; No. 50: 1343-44

23 Wilder, A.M., Allen, F.M., Power, M.H. and Robertson, H.E.: Carcinoma of the islands of the pancreas; hyperinsulinism and hypoglycemia.JAMA 1927; 89: 348-55

24 McGavran, M.H., Unger, R.H., Polk, H.C., et al. A glucagon-secreting alpha-cell carcinoma of the pancreas. N Engl J Med. 1966; 274:1408

25 Conn, J.W.: Aldosteronism and Hypertension. Primary Aldosteronism versus Hypertensive Disease with secondary aldosteronism. Arch Intern

Med: 1961 Jun; 107: 813-828

26   Conn, J.W.: Plasma Renin Activity in Primary Aldosteronism.JAMA
     1964; 190: 222-225
27   Conn, J.W., Cohen, E., Rovner, R.D.: Suppression of Plasma Renin
     Activity in Primary Aldosteronism: JAMA 1964; 190: 213-228
28   Conn, J.W., Cohen, E.L., Rovner, D.R. and Nesbit, R.M.: Normokalemic
     primary aldosteronism. JAMA 1965; 193: 200-205

12

Thyroid Cancer and the Atomic Age
The Chicago Experience

Raymon H. Grogan
Sharone Kaplan
Peter Angelos
Edwin L. Kaplan

The discovery of radiation and the atomic age have had far-reaching consequences, both positive and negative. The beneficial effects of radiation as a diagnostic and therapeutic agent are profound, while on the other hand the atomic age has led to enormous destruction and has cast a threat and danger upon the entire world. The University of Chicago played a key role in the development of the atomic age, as well as in the understanding of the deleterious effects of radiation on the thyroid. To understand how the University of Chicago and thyroid cancer fit into this global phenomenon we offer a broad overview of the history of the atomic age, and the consequences of radiation on the thyroid gland, with a particular focus on the Chicago Experience

In 1895 Karl Wilhelm Roentgen, a physics professor from Wurzburg, Germany, began experimenting with invisible cathode rays (Figure 1a).[1] He found that if these rays were projected onto a barium platinocyanide screen, they would fluoresce and could produce outlines of the bony structure of a human hand. One of the first photographic films he produced using this process was an image of his wife's hand (Figure 1b). Dr. Roentgen won the Nobel Prize in 1901 for this revolutionary work. The technique was clearly exciting and by 1896 radiographic films were being used for all types of diagnostic purposes. People quickly began experimenting with films of the long bones, chest, and even performed barium studies of the gastrointestinal system. It is interesting that even in these early days the damaging effects of radiation were noticed. Roentgen himself found that prolonged exposure to what he now termed "x-rays" could produce a wide array of negative side-effects, including skin ulcerations, hair loss, and dermatitis. But rather than focus on these negative aspects, researchers quickly adopted the new x-ray technology for

Figure 1a: Wilhelm Conrad Roentgen, the discoverer of x-rays, 1895. He received a Nobel Prize in Physics in 1901 for his discovery of X-rays.

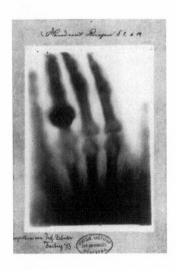

Figure 1b: First X-ray taken of Anna Roentgen's hand on December 22, 1895.

Figure 2: Maria Sklodowska Curie discovered radium. She received two Nobel Prizes for work on uranium, polonium, and radium.

therapeutic uses. Because it seemed that the x-rays caused mostly superficial damage, they were used with some success to treat a wide range of benign skin lesions including acne, hemangiomas, moles, and cutaneous tuberculosis. This newly discovered diagnostic and therapeutic tool was not just being used in Germany. Similar developments were occurring in other parts of Europe, and some of the most important work was taking place in Paris, France.

Maria Salomea Sklodowska, now more famously known as Madame Curie, left Warsaw, Poland in 1891 to study physics, mathematics, and chemistry at the Sorbonne in Paris, France (Figure 2).[2] She was 24 years old at the time. In 1895 she married her husband, Pierre Curie, also a physicist. Madame Curie studied uranium as part of a thesis project because her scientific mentor, Henri Becquerel, had recently discovered that uranium salts emitted rays that resembled x-rays in their ability to penetrate solid matter. Becquerel had discovered radioactivity, and he presented his findings from the uranium salts in 1896. Using hypotheses that were formulated by Maria, and a machine that detected radiation that was invented many years earlier by Pierre Curie, the husband and wife team expanded on Becquerel's initial work. By the end of 1898 they had discovered two new elements that they named "polonium" and "radium". By 1902 they were able to purify tiny quantities of radium chloride. In 1903 Madame Curie, Pierre Curie, and Henri Becquerel were awarded the Nobel Prize in Physics for their discovery of the phenomenon of radioactivity through their work on uranium. After the death of her husband in 1906 Madame Curie continued her work and by 1910 she was able to isolate the pure radium metal. In 1911 she was the sole winner of a second Nobel Prize, this time in Chemistry, for her work on polonium and radium. The discovery and purification of radium is what would ultimately lead to breakthroughs in radiation-based medical technology.

Just like Roentgen, the Curies quickly realized that radiation had both harmful and curative effects. Madame Curie sustained burns from carrying radium in her pocket, and in 1901 Pierre Curie proved the same phenomenon by purposely strapping radium to his arm. They also found that radium was 900 times more radioactive than uranium, and by 1904 Madame Curie had demonstrated that radiation therapy with radium could be used as a treatment for cancers deep in the human body. Unfortunately the full extent of the dangers of radioactivity would not be realized for another half-century. Madame Curie ironically died of either leukemia or aplastic anemia, probably from her long-term exposure to radiation.

Work conducted by the Curies stimulated immense excitement among

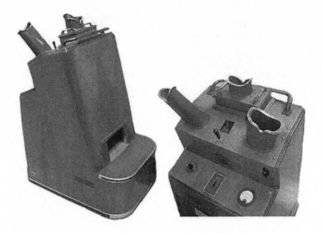

Figure 3: Fluoroscopes were commonly used in shoe stores to fit shoes properly from the 1930's to the 1950's. Very little attention was paid to safety.

physicians worldwide. By the time World War I started in 1914, radiation institutes had been opened all over the world. More than 100 diseases, both benign and malignant, were being treated with radiation therapy. Eventually enthusiasm outran caution and by the 1940's radiation usage had spread to alarming proportion. Fluoroscopes were being used casually in shoe stores to fit shoes (Figure 3). X-rays were being taken of fetuses still in the womb, and benign disorders like menstrual bleeding were being treated with radiation. By the early 1940's the long-term effects of repeated exposure to radiation were still unknown, or were being ignored, and many x-ray technicians as well as patients suffered the consequences (Figure 4). By this time radiation treatments had spread all across the globe, and had become well established in the United States. In fact, the University of Chicago became well known during the early 1940's for radiation treatment of benign peptic ulcer disease.

The University of Chicago was founded in 1890, coincidentally around the time that Roentgen and Madame Curie were starting their now famous experiments on x-rays and radioactivity. The physics department was founded shortly thereafter in 1893 and got off to a rapid and illustrious start. In 1907 University of Chicago physicist Albert Michelson became the first American to win the Nobel Prize. He won the Nobel Prize in physics for his work in optics and light waves. By 1939 researchers associated with the physics department had won a total of 8 Nobel Prizes. It is easy to understand why the

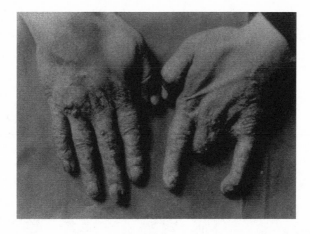

Figure 4a: Complications of x-rays were common in early radiologists and technicians including dermatitis, ulcerations, and squamous cell cancers. An amputation of the finger of this x-ray technician was done for a squamous cell cancer. This photo was taken circa 1910-1920.

physicists at the University of Chicago played a key role in the Manhattan Project, the research program the United States started in 1939 that would eventually lead to production of the first atomic bomb.

Perhaps the most famous University of Chicago physicist was the Italian-born Enrico Fermi. In 1938, at the age of 37, Fermi was awarded the Nobel Prize in physics for his work on nuclear reactions and his discovery of new radioactive elements. He fled the Mussolini regime that same year (his wife was Jewish) and emigrated to the United States. He started work immediately

Figure 4b: Ulceration of the skin was a complication of "beauty treatments" to the skin. These types of treatments were prevalent in the 1930's and 1940's in the United States.

Figure 5a:The first sustained nuclear chain reaction took place under the west stands of Stagg Field at the University of Chicago on December 2, 1942. A squash court with thick concrete lining under the stands of the football stadium was chosen as the site in naïve expectation that this would contain any damage if an explosion occurred.

at Columbia University and was an integral part of the first team of scientists to detect the energy released from a nuclear fission reaction. He soon moved his lab to the University of Chicago and was named the Associate Director of the Manhattan Project. On December 2, 1942, a group of dignitaries gathered with Enrico Fermi and his team under the concrete steps of Stagg Field, the football stadium at the University of Chicago.[3] The atmosphere in the room was tense. There, in an old concrete squash court at 3:36 pm, Dr. Fermi initiated the first artificial, self-sustaining nuclear reaction (Figure 5). A coded phone call was sent from Chicago to Washington, "The Italian navigator (referring to Fermi) has landed in the new world...the natives are friendly," meaning that the experiment was successful. This was a critical step needed for the large-scale production of weapons-grade plutonium, which would later be used in the bomb that was dropped on Nagasaki. The atomic bomb that exploded over Hiroshima was made of uranium. So began the atomic age.

Medical historian Roy Porter wrote, "Exposure to radiation and its consequences did not become a deep and long-lasting public concern until the consequences of the devastation of Hiroshima and Nagasaki by atom bombs at the close of World War II in 1945 became evident. Modern medicine has

Figure 5b: The director of the Manhattan Project at the University of Chicago was Enrico Fermi. He received a Nobel Prize in Physics in 1938 *"for his demonstrations of the existence of new radioactive elements produced by neutron irradiation, and for his related discovery of nuclear reactions brought about by slow neutrons".*

sometimes been so engrossed in its healing mission as to be cavalier about evaluating safety, benefits, and costs."[4] In 1955, following studies of survivors who received fallout from Hiroshima, Nagasaki, and the atom bomb tests in the Marshall Islands, investigators released the first documentation of the risks of exposure to *ionizing radiation* as a cause of benign and malignant thyroid tumors.[5] The devastating effects of the atomic bombs dropped on Japan by the United States seemed to have an eye-opening effect on the medical field regarding the dangers of radiation exposure. The University of Chicago would soon again be thrust into the national spotlight, this time for its role in discovering the harmful effects of *external-beam radiation* on the thyroid.

One of the first signs of trouble that drew attention to the long-term negative effects of external-beam radiation was a dramatic increase in the in-

Figure 5c: A sculpture, Nuclear Energy, by Henry Moore stands on the site of the "Chicago Pile" and was unveiled on December 2, 1967 at 3:36 PM, exactly 25 years after the first nuclear chain reaction.

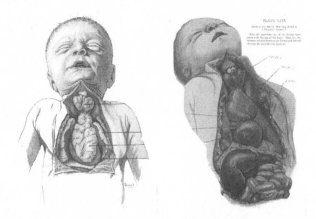

Figure 6: Infants who died a "thymic death." Sudden infant death was thought
to be due to compression of the trachea by thymic enlargement. From the first
edition of the book "Thyroid and Thymus", Andre Crotti 1918, published by Lea
& Febiger Philadelphia and New York. Left Figure 90 page 538, right Plate XXIX
opposite page 541.

cidence of thyroid cancer in children between 1930 and the early 1950's. At
the turn of the century physicians were treating large numbers of children and
adolescents with head and neck radiation for three major benign conditions:
an enlarged thymus, tonsillar disease, and acne. Thymic tracheostenosis, believed to be caused by an enlarged thymus, was considered to be a significant
contributing factor of sudden infant death syndrome.[6] It was characterized by
choking spells, dyspnea, asthma, cyanosis, and death. Status thymolymphati-
cus involving thymic hyperplasia, lymphatic hyperplasia, enlargement of the
spleen, reduced caliber of the aorta and great vessels, diminished resistance
to infection, and death was also a well-recognized disease in the early years of
the twentieth century (Figure 6). For acute cases, urgent surgical thymectomy
was imperative. For more chronic situations, irradiation to the chest, first described by Friedlander in 1907, was efficacious and therapeutic.[7]

In the 1920's, tonsillectomy was one of the most popular operations.[4] In
1925, for example, 25.5% of patients admitted to the Pennsylvania Hospital
in Philadelphia were suffering from diseases of the tonsils. In 1934 in New
York 1000 eleven-year-old children were examined for tonsillar diseases. Six
hundred and eleven children had already had a tonsillectomy. Of the other
389 children, after being examined by three panels of physicians, only 65 were
deemed healthy; tonsillectomy was recommended for the other 324 children.
However, this operative procedure was not without risk. Over 80 children in

England died each year from complications of a tonsillectomy. It is not surprising that radiation to the tonsils and adenoids became popular, starting in the 1930s. In Chicago, this treatment was advanced at Michael Reese Hospital as well as elsewhere.

In the early 1950s the medical community began to wonder why so many children were presenting with thyroid cancer. At the same time, the long-term effects of the atomic bombs were just starting to come to light. In 1950 Duffy and Fitzgerald at Memorial Hospital in New York published data suggesting that exposure to the low-dose external-beam radiation that was being used therapeutically world-wide might be a factor in the development of thyroid cancer in children and adolescents.[8] They found that 36% of thyroid cancer patients (10 of 28 patients) under age 18 who had been treated at Memorial Hospital had a history of radiation to the thymus gland between the fourth and sixteenth months of life. At the same time, University of Chicago physicians were busy conducting their own studies. Dwight Clark published the first Chicago case-series of 13 children with thyroid cancer under age 15.[9] Each of these children had been treated with radiation doses ranging from 200 to 725 rads to treat conditions including an enlarged thymus, cervical adenitis, enlarged adenoids and tonsils, sinusitis, peribronchitis, and pertussis. The latency period from time of radiation exposure to time of thyroid cancer diagnosis was 6.9 years.

Although Clark's study did not prove a definitive cause and effect, an association had been established. Other studies in the mid 1950s showed that female tuberculosis patients who were treated with radiation had an increased rate of breast cancer, and children born to mothers who had diagnostic x-rays while pregnant had high rates of leukemia.[10] By the early 1960s these data on therapeutic radiation combined with the data on ionizing radiation from the nuclear fallout victims made the risks of low-dose radiation therapy apparent.[11] The majority of radiation treatments for benign diseases were stopped. It was hoped that the problem with radiation-induced thyroid cancer in children would be a thing of the past. That was not the case.

In 1969, almost a decade after most people thought the issue of radiation-induced thyroid cancer had been resolved, Paul Harper and Daniel Paloyan reviewed the University of Chicago thyroid cancer cases from 1938 to 1967 and found 47 cases with a definite history of neck radiation.[12] The disturbing part of their report was that half of the patients had presented with thyroid cancer after age 20, with a mean latency period of 12 years after radiation exposure. Most reports on external beam radiation prior to this time had focused on thyroid cancer that developed in children, and thus most people assumed that thyroid cancer that presented in adults was not related to their prior radiation history. Although this study did not have a nationwide impact, it had the effect of starting a veritable explosion of research at the University of Chicago on the subject.

As follow-up on the Paloyan study, the following year Wilson, Platz, and Block reviewed the pathologic and clinical features of 58 cases of radiation-induced thyroid cancer at the University of Chicago during a 22-year period ending in 1968.[13] They found that most patients had been treated with neck radiation as children for thymus or tonsils, but 30% had radiation treatments as adults. Eighty percent were female, and the mean age at tumor diagnosis was 26.7 years. More than 80% were papillary cancers; local invasion was found in 60%, and lymph node metastasis was found in greater than 50%. Two deaths were due to thyroid cancer. Essentially, the Paloyan study pointed out that radiation-induced thyroid cancer had a latency period long enough to not be clinically apparent until adulthood. The follow-up Block study showed that neck radiation exposure in adulthood also conferred a risk for developing thyroid cancer.

In 1973 Leslie DeGroot and Edward Paloyan published a landmark paper in the Journal of the American Medical Association.[14] They reported that between 1968 and 1972 at the University of Chicago 50 additional patients were diagnosed with thyroid cancer, and 20 (40%) had a prior history of radiation to the neck. Most had been treated for tonsillar or adenoid disease (55%) when they were 3 to 7 years old or acne (25%) at age 15-20 years. Thirteen of the 20 patients were males, which was greatly different from the 21 females of 30 total patients in the non-irradiated group. The mean age at irradiation was 7.4 years and the mean age at diagnosis was 28 years. At this time Edwin Kaplan, who was appointed as a faculty member at the University of Chicago in 1968, performed the majority of thyroid cancer operations at the University of Chicago. Sixty-six percent of those operated upon with a thyroid mass and a history of radiation were found to have carcinoma.[14]

When knowledge of this study became known, it set off a small panic. Indeed, it was a deemed a Chicago Endemic.[14] Many medical students, physicians, and their families came to the university to be examined because they had had prior radiation exposure. DeGroot stated that the most important aspect of this study was to emphasize the *continued occurrence* of this problem. There were new radiation-induced cancers being found, but now in young adults. It was recommended that each person who had received low-dose radiation to the head and neck area should be examined and treated. The panic did not stay confined to the Chicago area for long, and soon it spread nationwide.

Following DeGroot's paper, a large registry of 5266 patients who had received external-beam radiation for tonsils and adenoids at Michael Reese Hospital in Chicago was found, and a recall program of these patients was initiated.[15] The recall program hit a nerve with the public. It was so well-publicized that the television program *Marcus Welby, MD*, featured the problem in a 1975 episode titled "The Time Bomb". Between 1973 and 1975, articles appeared in *Newsweek*, the *Wall Street Journal*, local papers and TV stations. It soon became apparent that this was not only a Chicago problem, but irradiated patients were being found throughout the entire Midwest and, in fact, throughout all the United States and Canada. Large recall programs were begun in Milwaukee and Detroit, and at other hospitals in Chicago. Soon the impact of these recalls was being reported.

The Michael Reese studies of the first 1476 patients who were examined revealed that 254 were operated upon and 92 carcinomas (36%) were found.[16] Chicago researchers projected that 10% of the entire group of these young patients who had been irradiated had thyroid caner. Cerletty and associates in Milwaukee examined 1825 subjects.[17] Nodular thyroid disease was considered an indication for operation. Of 110 patients operated upon, 32 (29.1%) of patients were found to have a carcinoma. In Detroit, Hamburger and Stoffer examined 814 patients with a radiation history in an 18-month period and found 16 new cancers.[18] In the previous 14 years they had only diagnosed 25 thyroid cancer patients who had a radiation history. DiGuilio et al examined 857 patients at five hospitals in the Detroit area during a one-year period.[19] Fifty-five patients had been operated upon and 12 cancers (22%) were found. Finally, at Evanston Hospital, north of Chicago, 530 patients were examined in an 18-month period; 60 patients were operated upon and 16 cancers (27%) were found.[20] Thus, in an 18-month period physicians identified and treated at least 175 cancers of the thyroid through the recall programs that had been initiated by the University of Chicago findings. Certainly, around the country hundreds of other cancers of the thyroid were identified in these groups of patients and, undoubtedly, many lives were saved. But the story was not over.

These previous studies had all been conducted on patients with a known or suspected history of thyroid cancer. In 1975, Refetoff at the University of Chicago reported in the *New England Journal of Medicine* on 100 patients with no known thyroid problems who were evaluated only because of a history of neck radiation.[21] Approximately 60% had had radiation for tonsils

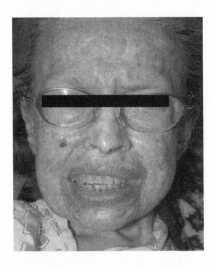

Figure 7: An 85-year-old female who was irradiated to her head and neck for acne as a teenager. Subsequently, she developed numerous skin cancers, a multinodular goiter, primary hyperparathyroidism, and cancer of the breast.

and adenoids. Twenty-six patients had a palpable abnormality and of 15 who were operated upon, seven were found to have a carcinoma. Thus, at least 7% of an unselected group who were examined only because of a history of neck radiation was found to harbor a thyroid carcinoma. All this accumulated evidence combined with public outcry led the U.S. Public Health Service to sponsor a national workshop and symposium concerning radiation-associated thyroid cancer. The conference was entitled "Radiation-Associated Thyroid Carcinoma" and was held at the University of Chicago from September 30 until October 1, 1976. DeGroot, Frohman, Kaplan, and Refetoff published the resulting proceedings in a 539-page document of the same name.[22]

As time has passed, it is evident that the Chicago Endemic is certainly over. After the near-hysteria of the early years, it has been clarified that radiation-induced thyroid cancer due to external radiation is *not* more aggressive than the usual papillary thyroid cancer and that it is very curable by standard techniques, including total (or near-total) thyroidectomy and radioiodine ablation.[23, 24] In 1990, DeGroot et al at the University of Chicago demonstrated that 100 irradiated thyroid cancers did not differ from 169 non-irradiated cancers in either mortality or virulence.[25] In 2009, Naing et al reported that of 4296 patients radiated at Michael Reese Hospital for benign conditions before age 16, 390 cancers (9%) were found but only 3 deaths occurred.[26] Physicians should be alerted to still ask about radiation exposure, for this is the most important historical fact that can be ascertained with respect to risk of thyroid cancer.

Unfortunately, this problem has not gone away. We continue to see patients at the University of Chicago who 50 years ago or longer had low-dose radiation to their neck.[27] In 2008 the United Nations issued a report that the worldwide radiation exposure was increasing secondary to diagnostic medical procedures. Reports have also been published implicating computed tomography scans as a source of increased lifetime cancer risk for individual patients. In 1986 many Eastern Europeans were exposed to the fallout from the Chernobyl nuclear accident as children or adolescents. It is estimated that the Chernobyl fallout will produce 16,000 cases of thyroid cancer by 2065. In 2011, the Fukushima Daiichi nuclear disaster once again focused the world's attention on radiation exposure in Japan.

The lessons learned from the story of radiation-induced thyroid cancer can be generalized to all of medicine. Physicians must not let enthusiasm outrun caution when it comes to incorporating new techniques or therapies into routine patient care. Surgeons in particular are often faced with difficult choices as new techniques are pioneered and introduced into mainstream practice. As physicians and scientists we must continue to be vigilant that none of our treatments, interventions, or discoveries that seem so helpful and beneficial today, will result in so much harm in the future (Figure 7).

References

1   Roentgen WC. Uber eine neue Art von Strahlen. Annalen der Physik und Chemi Neue folge, Vol. 64. Leipzig: Verlag Von Johann Ambrosius Barth; 1898:pp. 1-37.

2   Curie E, Sheean V. Madame Curie: A Biography. 2nd ed. New York: Da Capo Press, 2001.

3   Segrè E. Enrico Fermi: Physicist. Chicago: University of Chicago Press, 1970.

4   Porter R. The Greatest Benefit to Mankind: A Medical History of Humanity. New York: W. W. Norton & Company, 1997.

5   Cronkite EP, Bond VP, Conard RA, et al. Response of human beings accidentally exposed to significant fall-out radiation. *J Am Med Assoc* 1955; 159(5):430-4.

6   Crotti A. Thyroid and Thymus. Philadelphia and New York: Lea & Febiger, 1918.

7   Friedlander A. Status Lymphaticus and Enlargement of the Thymus: With Report of a Case Successfully Treated by the X-Ray. *Arch. Pediat.*

1907; 24:490-501.

8   Duffy BJ, Fitzgerald PJ. Cancer of the thyroid in children: a report of 28 cases. *J Clin Endocrinol Metab* 1950; 10(10):1296-1308.

9   Clark DE. Association of irradiation with cancer of the thyroid in children and adolescents. *J Am Med Assoc* 1955; 159(10):1007-9.

10   Macmahon B, Hutchison GB. Prenatal X-Ray And Childhood Cancer: A Review. *Acta Unio Int Contra Cancrum* 1964; 20:1172-4.

11   Kaplan EL. Hazards of "low dose" irradiation to the head and neck. *Curr Surg* 1978; 35(6):371-2.

12   Harper PV, Paloyan D. Thyroid carcinoma associated with radiation therapy. *Surg Clin North Am* 1969; 49(1):57-60.

13   Wilson SM, Platz C, Block GM. Thyroid carcinoma after irradiation. Characteristics and treatment. *Arch Surg* 1970; 100(4):330-7.

14   DeGroot L, Paloyan E. Thyroid carcinoma and radiation. A Chicago endemic. *JAMA* 1973; 225(5):487-91.

15   Favus MJ, Schneider AB, Stachura ME, et al. Thyroid cancer occurring as a late consequence of head-and-neck irradiation. Evaluation of 1056 patients. *N Engl J Med* 1976; 294(19):1019-25.

16   Frohman L, Schneider A, Favus M, et al. Thyroid carcinoma after head and neck evaluation of 1476 patients. In: Degroot L, Frohman L, Kaplan E, Refetoff S, eds. Radiation associated thyroid carcinoma. New York: Grune & Stratton; 1977:pp. 5-14.

17   Cerletty JM, Guansing AR, Engbring NH, et al. Radiation-related thyroid carcinoma. *Arch Surg* 1978; 113(9):1072-6.

18   Hamburger J, Stoffer S. Late thyroid sequelae of radiation therapy to the upper body. In: Degroot L, Frohman L, Kaplan E, Refetoff S, eds. Radiation associated thyroid carcinoma. New York; 1977:pp. 17-31.

19   DiGuilio W, Douglas R, Fink-Bennett D, et al. Results of screening patients with prior irradiation to the head and neck in five Detroit area hospitals. In: Degroot L, Frohman L, Kaplan E, Refetoff S, eds. Radiation associated thyroid carcinoma. New York: Grune & Stratton; 1977:pp. 33-34.

20   Murphy E, Scanlon E, Swelstad J, et al. A community hospital thyroid recall clinic for irradiated patients. In: Degroot L, Frohman L, Kaplan E, Refetoff S, eds. Radiation associated thyroid carcinoma New York: Grune & Stratton; 1977:pp. 35-40.

21   Refetoff S, Harrison J, Karanfilski BT, et al. Continuing occurrence of thyroid carcinoma after irradiation to the neck in infancy and childhood.

N Engl J Med 1975; 292(4):171-5.

22 DeGroot LJ. Radiation-associated thyroid carcinoma. A Grune & Stratton rapid manuscript production. New York: Grune & Stratton; 1977:pp. xx, 539 p.

23 Kaplan EL. Radiation-induced thyroid carcinoma. In: Najarian JS, Delaney JP, eds. Advances in breast and endocrine surgery. Chicago: Year Book Medical Publishers; 1986:pp. xiv, 535 p.

24 Calandra DB, Shah KH, Lawrence AM, et al. Total thyroidectomy in irradiated patients. A twenty-year experience in 206 patients. *Ann Surg* 1985; 202(3):356-60.

25 DeGroot LJ, Kaplan EL, McCormick M, et al. Natural history, treatment, and course of papillary thyroid carcinoma. *J Clin Endocrinol Metab* 1990; 71(2):414-24.

26 Naing S, Collins BJ, Schneider AB. Clinical behavior of radiation-induced thyroid cancer: factors related to recurrence. *Thyroid* 2009; 19(5):479-85.

27 Naunheim KS, Kaplan EL, Straus FHI, et al. High-does external radiation to the neck and subsequent thyroid carcinoma. Edinburgh ; New York: Churchill Livingstone, 1983.

13

The Founding of the American Association of Endocrine Surgeons
The Time was Right

Norman W. Thompson

By 1980, many advances had been made in the nascent discipline that was to become the field of endocrine surgery. From the mid 1950s, one discovery after another in both the clinical and basic sciences had been reported and a critical mass of information had been reached whereby the average general surgeon could no longer be expected to be up to date in the knowledge or clinical experience in the management of all of the surgical diseases involving the endocrine glands. Even the thyroid gland was a source of new discovery, with the classification of medullary carcinoma (MCT), identification of the C-cell as its precursor and calcitonin as its secretory product. An immunoassay for calcitonin had been developed that proved useful for the early detection of MCT in two new syndromes, MEN 2a and MEN 2b, in which the disease could occur in childhood. Subsequent to the 1955 report by Zollinger and Ellison, an explosion of interest in gut and islet-cell tumors and their hormonal secretions occurred. Gastrin was identified as the cause of the Z-E syndrome in the 1960s and other hormones were found as the cause of such rare conditions as the VIPoma and glucagonoma syndromes.

The discovery of immunoassays in the 1960s and '70s dramatically improved the ability to diagnose and treat a number of endocrine diseases at an earlier stage. Localization techniques including selective venous sampling, radioactive isotope imaging and CT scanning were applied to great advantage in managing endocrine diseases. The availability of cortisone in the late 1950s revolutionized the surgical management of Cushing's disease and syndrome and allowed for both pituitary and adrenal surgery to be performed safely. Pharmacological agents developed in the 1960s that included alpha- and beta-blocking drugs were used to prepare patients with pheochromocytomas for much safer operations. Hyperparathyroidism, considered an infrequent dis-

Figure 1a: The first AAES meeting in Ann Arbor, Michigan May 5ᵗʰ & 6ᵗʰ, 1980 with (left to right) Norm Thompson, Tony Edis, Jack Monchik, Orlo Clark, and Ed Kaplan

ease in the 1950s, became one of the most common endocrine surgical diseases. Routine serum calcium level measurements in the 1960s uncovered its frequency and PTH assays became available in the 1970s to confirm its presence. Multiple gland disease was found to be the most common manifestation of the MEN-1 syndrome that also involved the endocrine pancreas and the pituitary gland. Debates persisted as to how best to localize parathyroid adenomas, treat multiple gland disease, secondary and tertiary hyperparathyroidism and manage persistent and recurrent disease. Thus, there were many questions requiring answers, topics to be discussed and a forum greatly needed at which both could be accomplished.

Until 1980, there was no surgical society in the United States specifically focused on endocrine surgery. Although the American Thyroid Association (ATA) was founded in 1923 primarily by surgeons, by 1980 this organization had a limited surgical attendance and emphasized basic science investigation and medical aspects of thyroid disease. As early as 1974, during the ATA meeting in St. Louis, a small group of us expressed an interest in creating a society in which all aspects of endocrine surgery could be considered. At that time a more focused approach to endocrine surgery was being developed in Europe, particularly in Great Britain and Scandinavia. Several of the pioneers in the development of a strong multi-disciplined endocrine surgical program were at the Royal Postgraduate Medical School and Hammersmith Hospital. There, Selwyn Taylor and subsequently Richard Welbourn organized multidisciplinary researchers and clinicians to investigate all aspects of surgical endocrine diseases. Furthermore, Welbourn initiated a highly popular annual postgraduate course in endocrine diseases that attracted a large attendance from Europe as well as Britain. Several of our past AAES presidents were attracted to the Hammersmith for fellowship years to enhance their endocrine

Figure 1b: Auditorium in Towsley Center for Post-graduate Medicine at the University of Michigan where the first AAES meeting was held.

surgical training (including Orlo Clark and Richard Prinz). With some inspiration from events overseas and the steadily increasing volume of endocrine surgery, we established a Division of Endocrine Surgery at the University of Michigan in 1979, the first in the United States. This was soon to be followed at other centers. All patients with surgical disease of the thyroid, parathyroid glands, endocrine pancreas and adrenal glands were centered in one unit.

The American Association of Endocrine Surgeons (AAES) is indebted to some degree to the International Association of Endocrine Surgeons (IAES) that was founded on September 6, 1979 at the San Francisco Congress of the International Society of Surgeons (ISS). This new organization was the vision of Peter Heimann, a general surgeon and Professor of Surgery in Bergen, Norway. He had particular expertise and interest in the thyroid gland but was also interested in all other endocrine diseases. As early as 1976 he began to enlist interest in a new organization within the ISS for those surgeons who might have similar aspirations. When he unfortunately developed gastric carcinoma and died from this disease in March 1978, the task of organizing the first meeting was assumed by Selwyn Taylor and others with whom Peter had been corresponding. A one-day program was arranged for the proposed first meeting with scientific papers and panel discussions. The first business meeting establishing the IAES was held on September 6, 1979. Those elected were Selwyn Taylor as president, Richard Egdahl as president-elect, and Orlo Clark as secretary-treasurer. I served as coordinator of an eight-member executive committee. The aims of the IAES were "to provide a forum for the exchange of views of those involved in expanding the frontiers of endocrine surgery whether by clinical experience, laboratory investigation or in any other way. It was not for the general surgeon who occasionally operates on a thyroid or adrenal gland."

The seed for the AAES was planted that same day during a spontaneous luncheon at which those present, flush with the stimulation of participating in the founding of the IAES, proposed that an annual meeting of an American endocrine society might be endorsed by a sufficient number of surgeons to make this possible.

Those participating at that San Francisco luncheon included Orlo Clark, Tony Edis, Edwin Kaplan, Jack Monchik and myself. (Figures 1a and 1b) We were unanimously enthusiastic and optimistic that a meeting could be arranged for the spring of 1980 in Ann Arbor. A brief breakfast meeting on September 7th was held to solidify plans. It was decided that each of us would compile a list of all contemporary American surgeons who had made clinical or basic scientific contributions to endocrine surgery during the past ten years. The lists were to be circulated amongst the group, enlarged as appropriate and finalized by December, 1979. After more than one hundred candidate members were selected, a letter was sent to each proposing the new society and its goal of enhancing education, clinical care and research in the field of endocrine surgical disease. With a nearly unanimous affirmative response, a second letter of invitation was sent in January 1980 requesting abstracts and announcing a two-day scientific meeting to be held on May 5th and 6th in Ann Arbor at the University of Michigan Medical Center with me serving as chairman of local arrangements. The meeting was to start with a Sunday evening reception and dinner. Those present will always remember the enthusiastic response of "kindred spirits" joining in conversations that reached a very high decibel level. An immediate camaraderie developed that set the tone for this newly created group. The next two days of scientific meetings were exceptionally well-received and the ninety-six in attendance were in unanimous agreement to formalize as an organization initially called the American Endocrine Surgical Society. This name was changed to the American Association of Endocrine Surgeons after the first meeting. At our first business meeting, I was elected president, Orlo Clark vice-president and Jack Monchik secretary-treasurer. A 1981 two-day spring meeting was proposed for Washington, D.C. with Glenn Geelhoed to serve as chairman of local arrangements. The first meeting was considered to be very successful by all of those attending. There were a number of factors that contributed to this. The setting in the Towsley Center for Post-graduate Medicine was ideal and the tone was relatively informal with everyone encouraged to contribute... and they did. Finally, the program itself covered the entire spectrum of endocrine surgical diseases. The papers and discussions were excellent. For historical thoroughness an outline of the program follows.

The initial presentation was a brief "History of Endocrine Surgery" by Glenn Geelhoed. This was followed by a group of thyroid papers, which I chaired. A guest lecture was then given by Ronald Nishiyama, Director of Anatomical Pathology at Maine Medical Center in Portland, Maine. His talk was entitled "The Surgeon, the Pathologist and the Thyroid Gland." Subsequently, Ron Nishiyama was the first pathologist to be elected to honorary membership. He loyally attended and participated in all of our AAES meetings until illness prevented him from traveling. His talk was followed by a lively panel discussion on thyroid problems. The afternoon scientific session consisted of 7 parathyroid papers and was chaired by Orlo Clark. This was followed by a one and one-half hour panel discussion moderated by Orlo. The second day's program began with adrenal papers, chaired by Jack Monchik. This consisted of 6 papers considering aldosteronism, pheochromocytomas, paragangliomas, intravascular tumor extension and surgically correctible hypertension. A featured guest lecture was then presented by William Beierwaltes, Professor of Internal Medicine and the Director of the Section of Nuclear Medicine at the University of Michigan. His talk was entitled "Adrenal Scintigraphy: its present and future value to surgeons (Cortex and Medulla)." Just a few months later, in August 1980, the first pheochromocytoma was imaged with 131 I-MIBG developed in his laboratory. Beierwaltes had previously introduced NP-59, an iodo-cholesterol compound for cortical tumor identification. Following a business meeting and luncheon, the last scientific session consisted of pancreatic papers chaired by Ed Kaplan. The ensuing panel discussion, centered on islet-cell tumors of the pancreas, was moderated by Tony Edis.

It is noteworthy that many of those participating in presenting papers or panel discussions, subsequently became officers in the AAES and many attending the first meeting rarely missed another during the next twenty years.

Shortly after the meeting, I asked Jay Harness, then a young surgical faculty colleague at the University of Michigan, to draft a constitution for the AAES that could be circulated and voted upon at our next meeting. He efficiently completed this task and, the constitution still exists to this day with very few changes. Membership was to be limited to North American Surgeons with American Board Certification or its equivalence elsewhere who demonstrated a significant clinical or research interest in endocrine surgery. When overseas colleagues expressed an interest in attending our meetings, a Corresponding Membership category was established and overseas colleagues were welcomed to participate. Many have attended, contributed papers and discussion and en-riched the AAES with their presence. One of those, Tom Reeve from Sydney, Australia, was made an Honorary Member for his outstanding contribution to endocrine surgery. He is the only surgeon to be so honored by the AAES. Our close relationship with overseas colleagues culminated in a joint meeting in London and Lille, France, in 2000 that was organized by Jack Monchik and the British and French associations of endocrine surgeons.

Prior to the second meeting it was decided to have at least one guest lec-turer at each of our subsequent meetings to be selected by the president and local arrangements chair. A.G.E. Pearse from London was selected to lecture on APUDOMAS, his original concept, at the Washington D.C. meeting. This tradition has been maintained and many outstanding talks have been delivered to the AAES since then.

A firm decision was made after the second meeting to select papers on the basis of quality, whether clinical or basic sciences, providing they were origi-nal and not published or presented elsewhere. This was to ensure that there would be high level of quality for the scientific meeting and in anticipation that an excellent surgical journal might consider publishing our proceedings in the future. As a result, I contacted George Zuidema and Walter Ballinger, the two co-editors of *Surgery*, after our second meeting. Walter, who had attended the meeting, and George were both most enthusiastic and offered the entire December issue for AAES papers and discussions, providing sufficient quality papers were available. Initially one member of the AAES was to be appointed to the editorial board of surgery to facilitate this endeavor. Currently there are 8 members on the editorial board although only one officially represents the AAES.

In order to coordinate all of the activities associated with presentations, manuscript review and subsequent submission for publication, the office of Recorder was established in 1987 and Jon Van Heerden was the first to serve in that position. We were pleased that the discussions were to be published as well because they often contributed much to the presented paper. From the onset, there was no lack of participation by the membership. Unlike many other societies, invited discussants were never considered necessary. In fact, half of the audience has frequently lined up behind the available microphones and the moderators have often been required to curtail additional commen-tary. The congenial relationship with *Surgery* has persisted and the December issue continues to record the proceedings of the AAES meeting every year and is eagerly anticipated by the membership.

Our third president, Stanley Friesen from Kansas City, was the first member to receive the Oliver Cope Meritorious Achievement Award at the April meeting in 1994. This Award honors a member of the AAES in recognition of contributions in the field of endocrine surgery as investigator, teacher and clinical surgeon. A special award was first proposed to the council in April, 1984 and presented to Oliver Cope at the Toronto meeting in April, 1985. Cope was then Professor Emeritus at Harvard and considered America's foremost endocrine surgeon. His pioneering contributions to the development of parathyroid surgery were recognized worldwide. Afterward, the council decided that from then on the award would be known as the Oliver Cope Meritorious Achievement Award for the American Association of Endocrine Surgeons and to be given periodically to "members who truly aspire to this award." Stan Friesen who was a founding member and president from 1983-1984 was an outstanding investigator, educator and clinical surgeon who had a major interest in neuroendocrine tumors of the gut and pancreas as well as their associated hormones. In addition, he was a superb medical historian and fine musician (both concert and jazz piano.) He made many contributions with his wise counsel as well as his scientific reports until his death on February 28, 2008.

Within a year of the founding of the AAES, other national societies were being considered and the first to be organized and established was the British Association of Endocrine Surgeons with Dick Welbourn elected its first president. He was the Professor of Surgical Endocrinology at the Royal Postgraduate Medical School and Director of the Department of Surgery at the Hammersmith Hospital. Earlier in his career, while in Belfast in the 1950s, he was the first surgeon outside of the United States to successfully perform adrenalectomies in patients with Cushing's syndrome. In 1963, he authored the first major textbook in our field entitled *Clinical Endocrinology for Surgeons*. From 1983, after retiring as professor, he became a visiting scholar at UCLA and spent the next seven years researching the material for his book, *History of Endocrine Surgery*, published in 1990. Dick, who was clearly one of the early "spiritual" leaders in promoting and developing endocrine surgery, maintained close ties with many members of the AAES until his death in 2005.

After 30 years, the AAES has nearly quintupled its membership, including corresponding members from 24 countries outside of North America. Recently it has added Specialists in allied fields for membership and Resident / Fellow members to stimulate those in training and to enhance their experience in endocrine surgery. Endocrine Surgery has become a recognized surgical specialty and the AAES has been responsible for developing fellowships to assure that those choosing this field are adequately trained. The AAES continues to be a dynamic organization that has attained a national leadership role in promoting endocrine surgical patient care, education and scientific investigation. It has more than fulfilled the vision and aspirations of its founders.

A previous version of this chapter was published in *Surgery* (2011 Dec;150(6):1303-7). Reprinted with permission.

14

"Operating on Shadows"
The Discovery and Naming of Adrenal "Incidentaloma"

Wen T. Shen

Much of the modern endocrine surgeon's time is devoted to the workup and treatment of incidentally discovered tumors of the thyroid, adrenal, and pancreas. The concept of endocrine "incidentaloma" is firmly entrenched in clinical practice, and most surgeons have become used to the fact that many of the patients undergoing operations for these incidentally discovered tumors are asymptomatic and otherwise "healthy"; their only abnormality is a spot found on ultrasound, CT scan, or other imaging modality, which prompted further biochemical and/or cytologic workup and an eventual referral for resection. With the ongoing reliance of clinicians on imaging technology (when's the last time a patient with abdominal complaints made it through the emergency room without undergoing CT scan?), more and more studies are ordered each year, and with them comes a parallel increase in the number of incidentally discovered tumors. These tumors are a source of anxiety and confusion for clinicians and patients alike, and their workup and treatment levy an enormous cost to the U.S. health care system. Endocrine incidentalomas are so common in current practice, it is easy to forget that they are actually a relatively new clinical entity.

The term "incidentaloma" was coined in 1982 by George Washington University endocrine surgeon Glenn Geelhoed, and was initially used to describe an incidentally discovered tumor of the adrenal gland found on imaging study ordered for an unrelated indication.[1] Radiologists have commented on asymptomatic adrenal masses since the first X-ray images of the adrenal gland in the mid-20th century, but it was not until the introduction of CT scanning in the 1970s that adrenal incidentaloma became recognized as an unintended consequence of radiography and a potentially large-scale problem for clinicians. The first published reports of incidentally discovered adrenal tumors were from the early 1980s; two of these early reports, by endocrine surgeons Geelhoed and Richard Prinz, raised questions that would persist throughout the subsequent decades regarding the clinical significance and appropriate management of these tumors. Geelhoed and Prinz recognized even at this early stage that adrenal incidentalomas required clinicians to alter their traditional algorithms for diagnosing and treating adrenal disease, and thus generated uncertainty and anxiety regarding the decision to treat or observe them. Geelhoed was the first to use the word "incidentaloma" to describe an incidentally discovered adrenal mass; the origins of this new word, which has become an accepted part of medical terminology, shed light on the environment of confusion and uncertainty from which it was born.

The Italian anatomist Bartolomeo Eustachi provided the first anatomic descriptions of the adrenal glands in the 16th century, but the multiple hormonal functions of the adrenals were not delineated for another 300 years.[2] Tumors of the adrenal gland are found in approximately 3% of the population over 50 years of age in autopsy studies.[3] The majority of these adrenal tumors are neither malignant nor hormonally active, and are clinically silent. Functioning tumors of the adrenal gland are less commonly identified, and may be associated with clinically significant symptoms and signs such as hypertension, derangements of blood glucose and electrolytes, changes in body habitus, and psychological impairment. Tumors of the adrenal medulla that secrete excess catecholamines (pheochromocytoma) can cause life-threatening hypertensive crisis, cardiovascular collapse, and stroke. The adrenal gland may be a site of metastases from cancers of other organs, including lung, breast, kidney, and melanoma. Primary adrenal cancers are exceedingly uncommon, with 1-2 cases per million people in the U.S. diagnosed each year.[4]

Because the normal adrenal glands are small, surrounded by retroperitoneal fat, and in close proximity to other larger organs such as the liver, spleen, and pancreas, early radiographs did not typically identify them. Adrenal glands cannot be seen on plain X-rays unless they are calcified or quite enlarged. More advanced imaging techniques introduced in the mid-20th century gave radiologists slightly improved views of the adrenal glands; these techniques included excretory urography, contrast angiography, radioisotope scanning and ultrasound.[5] However, these modalities were still inadequate for identifying normal adrenal glands and most small adrenal tumors, and were dependent on the amount of contrast material administered, timing of the study, and experience level of the person performing and interpreting the study. In many instances the image of the adrenal gland obtained by these modalities offered only a suggestion or shadow of where the gland might reside. Radiologists during this period would occasionally comment on the "accidental" finding

of an adrenal mass during abdominal imaging, but these were rare occur-
rences that did not appear to merit much attention from radiologists or other
physicians.[6] Prior to the 1970s, adrenal tumors were not diagnosed until they
secreted enough hormones to generate signs and symptoms of hormonal excess, or grew large enough to become palpable on physical examination or
cause discomfort to the patient. A 1974 paper by Bernard Lewinsky from the
Royal Marsden Hospital in London detailed the clinical presentation of 178
patients with non-functioning adrenal tumors; nearly all of the patients exhibited a palpable mass or abdominal pain that prompted diagnostic evaluation
and treatment, and none of the patients was asymptomatic.[7]

The imaging modality that gave radiologists the first clear pictures of the
adrenal glands was computed tomography (CT). Invented over the course of
the 1960s and early 1970s through separate efforts by Allan Cormack of South
Africa and Godfrey Hounsfield of England (who were later awarded a joint
Nobel Prize in Medicine in 1979), the CT scanner uses X-rays to generate thin
cross-sectional images that can be formatted to create a three-dimensional re-
construction of the body.[8] The first commercially available CT scanners were
introduced between 1974 and 1976 and shortly thereafter radiologists began
utilizing this technology for identifying normal and diseased adrenal glands.
Nolan Karstaedt and colleagues of Washington University in St. Louis published a paper in 1978 describing their initial experiences with CT scanning of
the adrenal gland.[9] They first examined 200 "nonpathologic" abdominal scans
in order to determine how accurate the technique was in identifying normal
adrenal glands. Using different intervals between CT scan "slices," they were
able to identify 95% of normal adrenal glands, a tremendous improvement
over the less than 50% using prior modalities. This group then performed CT
scanning in 29 additional patients with known adrenal tumors and were able
to identify all of these tumors, some measuring 1 centimeter or smaller. While
the authors stated that CT scanning could demonstrate normal and enlarged
adrenal glands "safely, rapidly, and effectively," they urged caution in drawing
"definite conclusions" about "specific pathologic diagnosis" from the results
of CT scans and deferred to their colleagues in endocrinology to provide definitive identification of adrenal tumors based on biochemical tests and other
non-radiographic information.[9] At this early stage in the history of CT scanning, patients undergoing adrenal imaging were still expected to present with
"clinical and/or biochemical evidence of disturbance in adrenal function" or
"palpable upper-abdominal mass;" CT scan was used to confirm the clinical
diagnosis and to replace other more invasive radiographic methods. The au-

thors did mention the possibility of identifying incidental adrenal enlargement or "previously silent metastatic disease" on CT scan during "investigation of another suspected abnormality," but focused almost all of their attention on clinically detectable adrenal masses.[9] Within a few years, the typical algorithms for the workup of adrenal disease would be dramatically changed, and CT scan would become the primary means by which adrenal disease was diagnosed. As the number of abdominal CT scans and the experience of the radiologists interpreting them increased, physicians soon began to notice a parallel increase in the number of incidentally discovered adrenal masses.

Richard Prinz, an endocrine surgeon on faculty at the time at the Loyola University School of Medicine in Chicago, reported in 1982 on his institution's experience with nine asymptomatic patients who had incidentally discovered adrenal masses found on abdominal CT scan.[10] Presented at the third annual meeting of the American Association of Endocrine Surgeons in Houston, Texas on April 5, 1982, and published later that year in the *Journal of the American Medical Association,* this was one of the first papers to specifically investigate the problem of incidental adrenal mass found on CT scan. The nine patients described in the paper did not manifest any of the traditional signs and symptoms of adrenal hormone excess, although eight of the nine had hypertension. The tumors detected on CT scan ranged in size from 1 to 4 centimeters, representing minimal to modest increase in size from normal. All patients underwent biochemical testing for hormone excess, but only one had definitively elevated catecholamines (pheochromocytoma), a clear indication for operation. Because of the uncertain risk of malignancy, seven of the remaining eight patients were taken to the operating room for open adrenalectomy, a major abdominal operation requiring several days of hospitalization and significant postoperative pain. The final pathology results of the tumors removed included four benign cortical adenomas, two benign cysts, one lipoma (benign fatty tumor), and one pheochromocytoma (which had been suspected preoperatively by biochemical tests). There were therefore no cancers found in any of the operative specimens.

The subheading of Prinz's paper asked "Is Operation Required?": in retrospect, only one of the eight patients who underwent adrenalectomy actually benefited from surgery. Because of this low rate of hypersecreting or malignant tumors, Prinz urged that "care in interpreting the clinical significance of these masses and caution in recommending treatment are required."[10] Even at this early stage of recognition of these incidental adrenal tumors, Prinz was able to pinpoint the major source of tension that they raised:

> Does detection of these anatomic abnormalities offer hope for early diagno-
> sis and treatment of adrenal neoplasms so that complications of hormone-
> producing neoplasms can be lessened and the dismal prognosis of malignant
> tumors can be improved? Or does CT merely bring to clinical attention ad-
> renal enlargements that do not pose a threat to the patient's overall health?[10]

Prinz therefore recommended hormonal testing on all patients with incidentally discovered adrenal masses, along with a careful search for primary
cancers in other organs that could metastasize to the adrenal gland. If these
inquiries turned up no positive information, then he believed that "treatment
must be individualized,"[10] and offered a possible tumor size cutoff of 3 centimeters, or any growth on serial examinations as possibly indicating increased
risk of cancer. These guidelines would be refined and built upon by numerous other clinicians in subsequent years. In this influential 1982 article, Prinz
thus alerted the medical community to the new clinical entity of the incidentally discovered adrenal mass, and framed the question of how these tumors
should be perceived by clinicians: as harbingers of future malignancy or hormonal hyperactivity that merited aggressive treatment, or as inconsequential
radiographic findings that could be left alone? The incidentally discovered
adrenal mass would soon receive a catchier name, one that played upon the
uncertainty and confusion that it generated.

Glenn Geelhoed is an endocrine surgeon at the George Washington University Hospital in Washington, D.C. At the same 1982 meeting of the American Association of Endocrine Surgeons where Prinz had reported his initial
series of 9 patients with incidentally discovered adrenal masses, Geelhoed presented his own experience with the management of 20 similar patients.[1] The
resultant paper, co-authored with radiologist Edward Druy, was published
in *Surgery* later that year, a few months after Prinz's paper in *JAMA*. In his
presentation Geelhoed highlighted the fact that in previous decades, patients
with adrenal tumors almost always presented with signs and symptoms of
hormonal excess, and that "the inability of localization techniques to reveal
the anatomic source of functional abnormalities" was the major limiting step
in treating the problem.[1] The invention of abdominal CT scanning, however,
had created an "obverse problem": the "discovery of masses of unknown or
doubtful clinical significance."[1] CT scan had become the point of diagnosis

rather than the confirmation of clinical suspicion or positive lab tests for these patients. Geelhoed went on to present 20 patients with incidentally discovered adrenal masses, most of which were discovered on abdominal CT scanning. He detailed the clinical presentations of the patients, the radiographic appearances of the tumors, and the subsequent hormonal testing that was performed. Only one patient had biochemical abnormalities, and these results were equivocal at best. Nine patients ultimately underwent open adrenalectomy, almost "solely on the basis of the abnormal adrenal images."[1] Of these nine patients, six had benign pathology, mostly cysts or cortical adenomas; one of these patients had a completely normal adrenal gland. The remaining three patients included two with adrenal metastases from lung cancer, and one with an incorrect diagnosis who actually had a large retroperitoneal sarcoma that did not originate from the adrenal gland. Thus, no cases of hormone-secreting tumors and no cases of primary adrenal malignancy were identified from these 20 patients with incidentally discovered adrenal masses. Geelhoed stated that the radiographic findings of incidental adrenal tumors in these patients therefore "proved unnecessary and even harmful information in many of these patients."[1]

Geelhoed was not shy in stating his opinions regarding the potential hazards with identifying these clinically silent adrenal tumors. He urged caution in interpreting the information provided by radiographs in the absence of solid clinical evidence; otherwise surgeons would merely be "operating on shadows."[1] Since "the adrenal gland is no longer as hidden as it once was,"[1] clinicians would soon be facing increasing numbers of these incidental adrenal masses, and would need to remember that "the presence of an adrenal mass is not an indication for its removal."[1] As proven by Geelhoed's own experience of taking out multiple benign lesions with no primary adrenal cancers or hormonally active tumors to show for it, surgeons were in danger of subjecting patients to invasive operations with no clear benefit. Geelhoed was thus in a somewhat unusual position for a surgeon; rather than recommending intervention, he was serving as a self-proclaimed "protector of the adrenal gland"[11] against both unnecessary surgery as well as "the promiscuous use of adrenal imaging."[1] Echoing some of the themes raised in Prinz's *JAMA* paper, Geelhoed's paper brought to light a few of the most salient problems created by incidental adrenal tumors: the reversal of the conventional algorithms for diagnosis of adrenal tumors, the predilection of clinicians to seek action rather than inaction when faced with situations of uncertainty, and the rise to dominance of radiographic technology over clinical judgment.

Besides being one of the earliest case series of patients with incidentally discovered adrenal tumors, Geelhoed's 1982 paper continues to be widely cited today because it was the first published usage of the term "incidentaloma." Geelhoed did not explicitly define the term in his paper but did use it to describe any asymptomatic adrenal mass discovered on imaging studies ordered for another indication. The suffix "-oma" is of Greek origin and is used to denote a form of swelling or tumor. Numerous types of benign and malignant growths have names ending in "-oma," including carcinoma, lymphoma, sarcoma, and melanoma. Medical slang includes a few instances where "-oma" is added to an existing word to create a new entity: a "fascinoma" is a fascinating case or patient, a "horrendoma" is an unusually bad or gruesome case or patient. Geelhoed coined the term "incidentaloma" to describe an incidentally discovered adrenal tumor, and after the 1982 publication of this initial paper, the name stuck. Geelhoed now states that his initial creation of the word "incidentaloma" was intended to be sarcastic; by creating a somewhat funny name he was "belittling the diagnosis" and drawing attention to the fact that the majority of these tumors were not of any clinical importance.[11] In the paper he stated repeatedly that adrenal incidentalomas represented a "non-disease"[1] in the majority of cases, and that rational clinical decision-making should always trump a radiographic finding of unknown significance. Despite Geelhoed's intentions to use levity to help defuse some of the anxiety that these tumors would inevitably evoke in physicians and patients, the word "incidentaloma" would soon become an accepted term in standard medical practice, and its use would be expanded to include incidentally discovered masses in any organ. As Geelhoed now puts it, his simple turn of phrase "turned around to bite me in the butt."[11]

In the decade that followed Geelhoed's 1982 *Surgery* paper containing the first documented use of the word "incidentaloma," numerous papers on the subject of incidentally discovered adrenal masses were published. Most utilized or made reference to the newly coined term "incidentaloma;" Geelhoed's half-joking play on words had quickly become part of accepted medical jargon. Interestingly, several of these early papers using the word "incidentaloma" appeared in foreign language journals from both Europe and Asia, a sign that this subject was not just being encountered and analyzed in the U.S., and also an indication of the rapid transmission of medical information and terminology in the late 20th century. The word "incidentaloma" was also applied to incidentally discovered masses in organs other than the adrenal gland, including the pituitary gland, liver, and thyroid. In a 1989 letter to the editors of *Surgery*, Greek surgeon Dimitrios Linos criticized the rapidly growing usage of the word "incidentaloma";[12] Linos believed the name to be an inaccurate reflection of the tumor it was meant to describe, and felt that the original meaning behind the name was already being lost, as "incidentaloma" was being used to denote all types of adrenal tumors and not just those incidentally found in otherwise asymptomatic patients. While the name "incidentaloma" was admittedly "euphonic,"[12] Linos proposed that it be replaced by the more precise term "adrenaloma," which designated the organ of origin and highlighted the difficulty in determining whether the mass was benign or malignant. Linos' suggested name change never achieved widespread use (except by Linos himself, who published three more papers utilizing his preferred name); Geelhoed's original term, published in the same journal seven years prior, had already become ingrained in the medical vocabulary.

Abdominal CT scanning was introduced in the mid 1970s, and in less than a decade a new clinical entity born from this technology had been recognized and named. Endocrine surgeons Richard Prinz and Glenn Geelhoed, the first to publish articles on incidentally discovered adrenal masses, were quick to identify the potential problems that these tumors raised. They had already demonstrated that adrenal incidentalomas represented a departure from the traditional algorithms of diagnosis and treatment of adrenal disease, and had warned against the temptation to rush to operation for these tumors, given their exceedingly low overall rate of malignancy. In the years that followed, awareness of adrenal incidentaloma would spread throughout the medical community. Increasing numbers of patients underwent adrenalectomy for these incidentally discovered tumors (especially following the introduction of laparoscopic adrenalectomy in 1992[13]), with the majority demonstrating benign adrenal cortical adenoma on final pathology analysis. The clinical problems posed by adrenal incidentaloma were deemed significant enough to merit a 2002 NIH Consensus Conference that was solely devoted to the "Management of the Clinically Inapparent Adrenal Mass."[14] Prinz and Geelhoed's forewarnings would prove remarkably prescient; even today, 30 years later, endocrine surgeons continue to struggle with the vexing issues raised regarding observation versus operation for adrenal incidentalomas. There remains much room for improvement in the endocrine surgeon's role as "protector of the adrenal gland."

References

1   Geelhoed GW, Druy EM. Management of the adrenal "incidentaloma". *Surgery* 1982; 92(5):866-74.

2   Welbourn RB. The history of endocrine surgery. New York: Praeger, 1990.

3   Mansmann G, Lau J, Balk E, et al. The clinically inapparent adrenal mass: update in diagnosis and management. *Endocr Rev* 2004; 25(2):309-40.

4   Shen WT, Sturgeon C, Duh QY. From incidentaloma to adrenocortical carcinoma: the surgical management of adrenal tumors. *J Surg Oncol* 2005; 89(3):186-92.

5   Ferris EJ, Seibert JJ. *Urinary Tract and Adrenal Glands: Multiple Imaging Procedures.* New York: Grune & Stratton, 1980.

6   Rigler LG. *Outline of Roentgen Diagnosis.* 2nd ed. Philadelphia: J. B. Lippincott Company, 1943.

7   Lewinsky BS, Grigor KM, Symington T, et al. The clinical and pathologic features of "non-hormonal" adrenocortical tumors. Report of twenty new cases and review of the literature. *Cancer* 1974; 33(3):778-90.

8   Kevles B. Naked to the bone : medical imaging in the twentieth century. New Brunswick, N.J.: Rutgers University Press, 1997.

9   Karstaedt N, Sagel SS, Stanley RJ, et al. Computed tomography of the adrenal gland. *Radiology* 1978; 129(3):723-30.

10 Prinz RA, Brooks MH, Churchill R, et al. Incidental asymptomatic adrenal masses detected by computed tomographic scanning. Is operation required? *JAMA* 1982; 248(6):701-4.

11 Geelhoed G. telephone interview with author. June 16, 2008.

12 Linos DA. Adrenaloma: a better term than incidentaloma. *Surgery* 1989; 105(3):456.

13 Gagner M, Lacroix A, Bolte E. Laparoscopic adrenalectomy in Cushing's syndrome and pheochromocytoma. *N Engl J Med* 1992; 327(14):1033.

14 National Institutes of Health. Program and Abstracts from the NIH State-of the Science Conference on the Management of the Clinically Inapparent Adrenal Mass ("Incidentaloma"). *In* National Institutes of Health, ed., Vol. 3. Bethesda, 2002.

15

The Discovery of the RET Proto-Oncogene and its Role in MEN-2

Geoffrey W. Krampitz and Jeffrey A. Norton

Introduction

Advances in genomic technologies marked by the sequencing of the human genome and accelerating improvements in molecular diagnostics have ushered in an era of increasingly personalized medicine. Such an approach employs the genomic, proteomic, and molecular information unique to the individual. This information can then be used to provide an accurate genetic diagnosis, molecular risk assessment, informed family counseling, therapeutic profiling, and early preventive management that best fits the individualized needs of each patient. Identifying mutations in the RET proto-oncogene as the causative factor in multiple endocrine neoplasia type 2 (MEN 2) has led to the development of direct genetic testing for at-risk individuals. Patients with germline RET mutations may undergo risk assessment based on particular mutations and may be offered thyroidectomy prior to developing clinically evident medullary thyroid cancer (MTC) or early in the course of the disease when cancer is limited to the thyroid gland. Family members may also receive counseling based on understanding of the genetic transmission of the disease. We are now embarking on an era of increasingly personalized medicine where therapeutic choices may be guided by an informative biomarker 'code.'[1] The revolutionary changes in detection and management of MEN 2 ushered by linking RET mutations and the onset and severity of the syndrome are some of the clearest examples of the benefits of personalized medicine. In this chapter, we describe the historical context of the discovery of the RET proto-oncogene and its role in the diagnosis and treatment of MEN 2.

MEN 2

Multiple endocrine neoplasia type 2 is composed of three distinct clinical sub-types, MEN 2a, MEN 2b, and familial medullary thyroid carcinoma (FMTC). MEN 2 is a rare syndrome with an incidence of 1 in 200,000 live births. Each subtype is an autosomal dominant familial cancer syndrome associated with germline mutation of variable penetrance in the RET proto-oncogene[2] Because 50% of offspring of an affected parent will manifest MEN 2, the syndrome tends to occur in every generation of a family. John Sipple at the State University of New York in Syracuse recognized the hereditary nature of the syndrome in 1961. It was during this time that the mechanisms responsible for genetic transmission were being deciphered. In 1952, Hershey and Chase confirmed the earlier findings of Avery, MacLeod, and McCarty that DNA is the molecule responsible for mediating heredity. The contributions by Franklin and Wilkins led to the revelation of the helical structure of DNA by Watson and Crick in 1953. In 1959, messenger RNA was found to be the intermediate between DNA and protein, and the genetic code governing translation was subsequently elucidated in 1966. Thus, the discovery of MEN 2 as a hereditary syndrome occurred during the dawn of the genomics era when understanding the genetic basis of diseases was becoming possible. Consequently, a major impetus in the scientific work that followed the discovery of MEN 2 involved deciphering the gene responsible for the syndrome and realizing the promises of such breakthroughs.

The principal feature of all MEN 2 subtypes is medullary thyroid carcinoma (MTC), a cancer of the parafollicular (C) cells. Patients with MEN 2 have a 70-100% risk of developing MTC by age 70 years, and MTC is the most important determinant of mortality in these patients. MTC has a poor prognosis once it has spread beyond the thyroid. However, patient survival is greatly improved when MTC is treated early in the course of the disease. Therefore, much of the emphasis of treating patients with MEN 2 has focused on identifying and treating MTC.[2]

In 1980, Glen W. Sizemore at the Mayo Clinic in Rochester, Minnesota, recognized that MEN 2 was composed of two variants that he called MEN 2a and MEN 2b.[3] Common features of the two groups are multicentric MTC, occurring in greater than 90% of patients, and bilateral pheochromocytomas, occurring in 50% of patients.[4]

MEN 2a is the most common manifestation of MEN 2, accounting for

55% of cases. MTC is often the first manifestation of MEN 2a, usually occurring between the ages of 20–30 years. MEN 2a is characterized by MTC and pheochromocytomas plus primary hyperparathyroidism, manifesting in 20-30% of patients.[2]

MEN 2b is the rarest of the variants, accounting for approximately 5-10% of all cases. MEN 2b is marked by earlier onset of disease compared to other forms of MEN 2. Usually MTC develops within the first year of life, and the clinical course is usually more aggressive due to more advanced disease at presentation.[5] MEN 2b lacks parathyroid pathology, but instead features Marfanoid habitus, mesodermal abnormalities, labial and mucosal neuromas, and gastrointestinal ganglioneuromas, often leading to constipation and megacolon.[2-4, 6]

In 1986, Farndon at the University of Newcastle upon Tyne, United Kingdom, and Samuel Wells at Washington University School of Medicine in Saint Louis, Missouri, described a third variant, FMTC. This variant is the second most common, accounting for 35% of total cases. FMTC is the mildest subtype and features MTC without any of the extra-thyroid characteristics of the other MEN 2 subtypes.[7] The diagnosis of FMTC should be considered when at least four family members develop isolated MTC.

MEN 2a

In 1961, John Sipple first described the association of large malignant tumors of the thyroid, bilateral adrenal pheochromocytomas, and parathyroid hyperplasia. In this report, he recalled the case of a 33 year-old man with a history of recurrent headaches who presented to the Veterans Administration Hospital in Syracuse, New York, with severe right-sided headache, hypertension, nausea, and vomiting followed by left-sided weakness and lethargy. A lumbar puncture on admission revealed bloody spinal fluid and an elevated opening pressure. A right carotid arteriogram revealed a large arteriovenous malformation. The patient underwent a craniotomy to evacuate an intracerebral hematoma and coagulate the network of coiled vessels. His postoperative course was marked by fever, restlessness, and fluctuating blood pressure. Seven days after his initial operation, the patient developed signs of increased intracranial pressure and required a second craniotomy to evacuate another hematoma. However, the patient died shortly after the operation. An autopsy revealed intracerebral hemorrhage, bilateral adrenal pheochromocytoma, and thyroid adenocarcinoma. The association of adrenal phochromocytoma and thyroid cancer intrigued Sipple, prompting him to conduct a retrospective analysis of similar cases in the literature. From a population of 537 patients with pheochromocytoma, he identified 5 other patients who also had thyroid cancer. He concluded that the incidence of carcinoma of the thyroid gland in these patients with pheochromocytoma was increased far beyond expectation based on chance concurrence.[8] Two years later in 1963, Preston Manning at the Mayo Clinic reported a case of recurrent bilateral familial pheochromocytomas, hyperparathyroidism with multiple parathyroid adenomas, as well as occult bilateral thyroid carcinoma containing amyloid stroma. The case involved an 18 year-old woman first admitted with palpatations, blurry vision, and headaches. On exam, the patient was hypertensive, pale, and with reduced visual acuity and papilledema. Pharmacologic tests with phentolamine supported the diagnosis of pheochromocytoma for which she underwent a surgical resection with resolution of her hypertension. Ten years later, she presented with "vertiginous attacks and burning parasthesias of the face." She was found to be hypertensive on exam. A detailed history revealed that in the interim she had pre-eclampsia during her third pregnancy, her brother had bilateral pheochromocytomas removed, and her father had died of a "Bright's disease" and a cerebrovascular accident. Laboratory tests revealed recurrent pheochromocytoma and hypercalcemia. In addition, an excretory urogram revealed bilateral nephrocalcinosis. She underwent operative exploration that revealed multiple recurrent pheochromocytomas in the right adrenal gland. The operation was complicated by catecholamine surge leading to cardiac arrest requiring transthoracic cardiac massage to restore normal cardiac rhythm. However, the tumors were removed successfully, and the patient recovered. One month later, the patient underwent cervical exploration that revealed three-gland parathyroid hyperplasia as well as thyroid carcinoma. The involved parathyroids were removed, and the patient underwent subtotal thyroidectomy and lymph node dissection. Postoperatively, the patient developed hypocalcemia, hypothyroidism, and adrenal insufficiency all of which were managed medically and the patient recovered well. Manning compared this case to a similar case at the Mayo Clinic of a 14 year-old girl who underwent a total thyroidectomy in 1952 for a histologically similar thyroid carcinoma. The patient later died of complications of placenta previa during pregnancy. However, on autopsy, she was found to have metastatic thyroid carcinoma as well as bilateral pheochromocytomas. Given the multiple neoplasms of endocrine organs in these cases, Manning suggested a relation to multiple endocrine adenoma syndrome (MEA), a syndrome affecting the pituitary, pancreas, and parathyroids described earlier by Jacob Erdheim in 1903, Laurentius Underdahl in 1953[9], and Paul Wermer in 1954. However, given the differences in the organs involved, Manning concluded that the patient represented a further variant of the MEA syndrome.[10] In 1968, Alton Steiner at the Albany Medical Center in Albany, New York, meticulously described a kindred of 168 patients, 25 of whom had pheochromocytomas (10 proven and 15 probable), 5 had MTC, and 2 had parathyroid chief cell hyperplasia. He concluded that the syndrome was genetically transmitted in an autosomal dominant fashion with high penetrance similar to MEA syndrome. However, because MTC and pheochromocytomas were defining features of this new syndrome, Steiner concluded that MEN 2 must represent a genetically distinct entity from MEA, which he proposed be renamed MEN 1.[11] Wells had several kindreds with MEN2a. He realized that most of the elders in the families died of metastatic MTC. In 1975 he worked on a method for the early diagnosis of MTC so that young offspring could possibly be cured of this inherited cancer. Studies showed that calcitonin (thyrocalcitonin) was an excellent marker for MTC in that the C-cells stained positive for calcitonin by immunohistochemistry and that patients with metastatic MTC had elevated serum levels of calcitonin. Studies in pigs showed that both calcium and pentagastrin stimulated C-cells to secrete calcitonin levels and could be accurately measured in the blood by radioimmunoassay. Wells studied several kindreds with MEN2a and measured calcium and pentagastrin stimulated levels of calcitonin in the peripheral blood. He was able to diagnose MTC when it was surgically curable and localized to the thyroid gland without lymph node metastases. Many patients were cured by this method, but some still had metastatic disease on final pathological analysis.

MEN 2b

In 1966, E. D. Williams and D. J. Pollock at the London Hospital reported two cases of neuromatosis, pheochromocytomas, and thyroid carcinoma. The first case involved a 33 year-old woman who presented with abdominal discomfort and blanching fingers. On exam, she had pedunculated nodules on the eyelids, lips, and tongue. There was a 3 cm nodule in the left lower pole of the thyroid that was surgically resected and found to be medullary carcinoma on histological analysis. Seventeen months later, she presented again with abdominal pain and was found to be hypertensive on physical exam. Ultimately, she was diagnosed with pheochromocytoma for which she underwent a left adrenalectomy revealing multiple tumors. Biopsies of the tongue and lip lesions showed plexiform neuromas. The second case was found during a study of the autopsy records of a 19 year-old woman with pheochromocytoma. The patient presented with attacks of choking sensations, headaches, paresthesias in the extremities, diarrhea, and profuse sweating. Family history revealed that her father shared the patient's coarse facial features, suffered from palpitations and flushing, and died at age 38 after an abdominal operation. On examination, the patient had thickened lips, multiple nodules along the border of the tongue and left lower eyelid, and a rounded mass in the right hypochondrium. Biopsies of the tongue and eyelid showed neuromata. On laparotomy, diffuse colonic thickening was found, along with a large mass in the upper pole of the right kidney from which bloody fluid was aspirated. Postoperatively, the patient became tachycardic and hypertensive and died shortly thereafter. An autopsy revealed a hemorrhagic right adrenal pheochromocytoma, multiple mucosal and ocular neuromas, ganglioneuromatosis of the intestine, and medullary carcinoma of the thyroid metastatic to the cervical lymph nodes. These two cases involved cutaneous lesions and pheochromocytomas akin to von Recklinghausen's disease. However, the histological findings of neuromas instead of neurofibromas, absence of extensive cutaneous involvement, and thyroid medullary carcinoma were inconsistent with von Recklinghausen's disease. Thus, Williams and Pollock concluded that the patients they described represented a syndrome allied but distinct from von Recklingausen's disease.[12]

In 1968, R. Neil Schimke at Johns Hopkins Hospital in Baltimore, Maryland, described three patients with bilateral MTC, bilateral pheochromocytomas, multiple mucosal neuromas, Marfanoid body habitus, coarse facial features, and characteristic skeletal abnormalities (pectus carinatum, saber shin). The first case involved a 53 year-old woman who was admitted for thyroid evaluation. Sixteen years earlier, she underwent a subtotal thyroidectomy for a goiter of unknown histology. Five years later she developed "glands in the neck," but no biopsy was performed at that time. She was hospitalized two years later out of concern for acromegaly given her unusual facial appearance. She was noted to have significant cervical lymphadenopathy for which she underwent a total thyroidectomy with neck dissection that revealed extensive tumor invasion into the trachea, esophagus, and carotid sheath. She was readmitted seven years later with abdominal distension and hypocalcemia. On exam she had "floppy" ears and a prognathic jaw, bilateral cataracts, thickened upper eyelid margins, numerous small nodules on the hard palate and tongue, and irregular masses in the cervical neck, kyphotic chest, and abdominal distension. Barium enema revealed megacolon, although rectal biopsy was normal. Cervical lymph node biopsy demonstrated metastatic MTC, and biopsy of a mucosal nodule revealed neurofibroma. The patient was treated medically with desiccated thyroid and antispasmotics without any significant change in symptoms. The second case involved a 25 year-old gentleman admitted for thyroid carcinoma and megacolon. Eight years earlier, he underwent a thyroidectomy for a malignant thyroid nodule and subsequently treated with radioactive iodine and irradiation. Over the next few years, he had nausea, epigastric pain, diarrhea, and weight loss. Barium enema showed a markedly dilated colon, although ganglion cells were present on rectal biopsy. In the months preceding admission, he developed chest pain, weakness, extremity swelling, and nocturuia. Physical exam revealed a tall, thin man with enlarged lips, thickened eyelids and scalp, prominent orbital ridges, broad nasal root, scattered yellow thickenings of the palpebral and bulbar conjunctivas, prominent corneal nerves, and extensive hyperpigmented papules on the skin, multiple firm cervical nodes, dorsal kyphosis with long, thin extremities. Barium enema again showed dilation of the entire colon. Cervical biopsies demonstrated metastatic MTC, and conjunctival biopsy showed neurofibroma. The patient was treated symptomatically and died shortly after discharge. An autopsy revealed diffusely metastatic MTC, bilateral pheochromocytomas, and intestinal intestinal ganglioneuromatosis. The third case involved a 23 year-old woman who presented with nausea, diarrhea, weight loss, and palpitations. Physical exam revealed diffuse hyperpigmented papules, small nodules throughout the oral cavity, and an enlarged and nodular thyroid. She underwent a subtotal thyroidectomy that revealed MTC, and was subsequently treated with external radiation and desiccated thyroid. Four years later, her symptoms recurred. She was diagnosed with a nonfunctioning left kidney due to pyelonephritis for which she underwent a nephrectomy. Postoperatively, the patient developed severe hypertension with laboratory test indicative of pheochromocytoma. However, the patient refused further surgery. Six months later, the patient was readmitted in shock and died shortly thereafter. An autopsy revealed bilateral pheochromocytomas, metastatic MTC, and neurofibromas of the Auerback and Meissner plexuses of the gastrointestinal tract. Through careful study of these cases, Schimke noted several unifying characteristics of the disorder with therapeutic implications. The syndrome was marked by bilateral, and often extra-adrenal, pheochromocytomas, multicentric MTC requiring total thyroidectomy and possibly lymph node dissection, parathyroid hyperplasia with hypercalcemia and nephrocalcinosis, neurofibraomas that may cause considerable autonomic dysfunction, and a familial mode of transmission. He correctly postulated that the entire syndrome could be explained on the basis of a heritable defect in a single cellular system within neural crest derived tissues.[13]

Familial MTC

FMTC is the mildest variant of MEN 2 and accounts for 35% of total cases. There is a strong predisposition for MTC, but not for other clinical manifestations of MEN 2. FMTC was first described in 1986 by Farndon and Wells.[7] They evaluated 213 patients from 15 kindreds with familial MTC. MTC was confirmed by plasma calcitonin with or without pentagastrin stimulation and treated with total thyroidectomy with central lymph node dissection. Evaluation for hyperparathyroidism included measuring total serum calcium, parathyroid hormone, phosphorus, chloride, and alkaline phosphatase. Pheochromocytoma was diagnosed by measuring 24-hour urinary excretion rates of adrenaline, noradrenaline, vanillylmandelic acid, metanephrine, and computerized tomographic scans in selected patients. There was no biochemical or radiographic evidence of pheochromocytomas, nor was there biochemical, macroscopic or microscopic evidence of hyperparathyroidism in any individual studied. In their analysis, Farndon and Wells detected 41 subjects from two kindreds who had clinically occult MTC but lacked extra-thyroidal manifestations of other MEN subtypes. Thirty-seven patients with MTC underwent total thyroidectomy, and four refused surgical intervention. No one within the kindreds analyzed died from MTC. FMTC patients were compared with MEN 2a patients matched for MTC tumor burden. The mean age of diagnosis of MTC in the FMTC group was 43 years compared to 21 years in the MEN 2a group. Thus, the clinical course of MTC in FMTC is less severe than in other MEN variants, and the onset of the disease is generally later in life.[14]

RET proto-oncogene

RET is a proto-oncogene composed of 21 exons located on chromosome 10 (10q11.2) and encodes for a transmembrane receptor tyrosine kinase for members of the glial cell line-derived neurotrophic factor family (GDNF) and associated ligands (artemin, neuturin, persephin).[15-18] In 1985, Takahashi at the Dana-Farber Cancer Institute in Boston, Massachusetts, screened for possible genes capable of transforming NIH 3T3 fibroblasts using a transfection assay that was used to previously identify other transforming genes, such as *ras, Blym-1, Tlym-I, and met.* Using sonicated human lymphoma DNA segments to

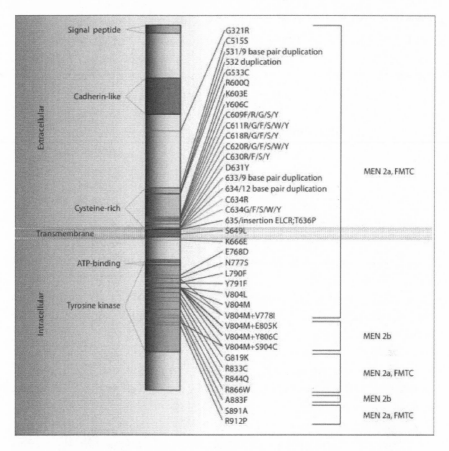

Figure 1. The RET protein is composed of three functional domains, including an extracellular ligand-binding domain, a transmembrane domain, and a cytoplasmic tyrosine kinase domain. The extracellular domain contains a signal peptide that is cleaved, four cadherin-like repeats,, and a cysteine-rich region critical for disulfide bond formation needed dimerization.

transfect fibroblast cells, Takahashi was able to transform these cells. Subsequent cloning and homology analysis suggested that the new transforming gene was unlike previously identified transforming genes. Using blot hybridization analysis, he determined that the transforming sequence encompassed 34 kb and represented a rearrangement of two unlinked human DNA segments. Molecular cloning indicated that the biologically active gene was generated by recombination of two separate normal DNA segments. Northern blot analysis indicated that the new gene produced a new transcriptional unit.

Because the new transforming gene was the product of gene rearrangement during transfection of 3T3 fibroblasts, the gene was named "REarranged during Transfection," or RET.[19] At that time, the clinical significance of the newly discovered RET gene was unknown. Thus, the search for the location, structure, and function of this gene in humans would soon commence.

The RET protein is composed of three functional domains, including an extracellular ligand-binding domain, a transmembrane domain, and a cytoplasmic tyrosine kinase domain. The extracellular domain contains four cadherin-like repeats and a cysteine-rich region. The highly conserved cysteine-rich region is important for disulfide bond formation needed for maintaining the native tertiary structure allowing for receptor dimerization. The intracellular domain contains two tyrosine kinase subdomains that are involved in several intracellular signal transduction pathways.[20] Figure 1 illustrates the RET protein domains highlighted with known mutations that are grouped by MEN subtype. Coupling RET, its co-receptors, GNDF-family receptor alpha (GFRα 1-4)[21], and GNDF-family ligands leads to RET dimerization to form a heterohexamer complex that results in transcellular kinase activation and signaling.

RET is involved in a number of cellular signaling pathways during development regulating the survival, proliferation, differentiation, and migration of the enteric nervous system progenitor cells, as well as the survival and regeneration of neural and kidney cells.[22-29] Mice homozygous for a targeted mutation in RET demonstrate renal agenesis or severe dysgenesis and lack enteric neurons throughout the digestive tract suggesting that RET is a component of a signaling pathway required for renal organogenesis and enteric neurogenesis.[30, 31] Subsequently, inactivating mutations in RET associated with Hirschprung's disease[32-37] and renal agenesis were identified.[30, 31, 38-42]

In 1987, genetic linkage analysis mapped the MEN 2a locus to a region on chromosome 10.[43, 44] In subsequent years, the location of the MEN 2a locus was further refined. In 1990, Massimo Santoro at the Institute of Experimental Endocrinology and Oncology in Naples, Italy, found increased expression of the RET gene in both familial and sporadic human pheochromocytomas and MTC. Because the RET gene was mapped to a location on chromosome 10 in close proximity to the gene that predisposes patients to the MEN 2a syndrome, he suggested that this region of chromosome 10 might be involved in the proliferative and differentiation of neuroectodermal tissues.[45] Wells knew that even pentagastrin and calcium stimulated serum levels of calcitonin failed to reliably diagnose patients with curable familial forms of MTC so he decided to try to identify the exact gene defect responsible for these syndromes. Furthermore, he realized that family members without the inherited mutation could be told that they did not have the syndrome and that they would not need serial screening as had been done with the pentagastrin-stimulated calcitonin levels. In 1993, Hellen Donis-Keller, working with Wells at Washington University School of Medicine, identified seven RET mutations in exons 7 and 8 associated with MEN 2a and FMTC. They screened a panel of genomic DNA from seven MEN 2a, MEN 2b, and FMTC families and normal controls. Using PCR and sequence analysis approaches, a total of seven mobility shift variants (SSCVs) were observed in MEN 2 families and not in DNA from normal individuals. Six of these were identified in exon 7 and one in exon 8, and all mutations affected codons specifying cysteine residues. SSCVs present in germline DNA were also found in MTC and pheochromocytoma from familial cases. However, these mutations were not found in three families with MEN 2a, suggesting that these were not the only mutations responsible for MEN 2.[46] That same year, Lois Mulligan at the University of Cambridge, United Kingdom, generated cDNA from four MTCs and three pheochromocytomas from patients with MEN 2a and sporadic tumors. Using chemical cleavage mismatch procedures and sequencing, she identified specific missense mutations in the RET gene at codons encoding for cysteine residues within the transition point between the extracellular and transmembrane domains. These mutations were found in 20 of 23 apparently distinct MEN 2a families, but not in 23 normal controls. Furthermore, 19 of these 20 mutations affected the same conserved cysteine residue at the boundary of the RET extracellular and transmembrane domains.[47] Subsequently, Robert Hofstra at the University of Groningen, the Netherlands, showed that MEN 2b is associated with a T644M mutation affecting the intracellular tyrosine kinase domain of the RET proto-oncogene. Using single-strand conformational polymorphism analysis, he found a variant pattern in all nine unrelated MEN 2b patients and six out of 18 sporadic MTC in his cohort.[48] In 1995, Santoro demonstrated the RET mutations responsible for MEN 2a and MEN 2b acted by constitutively activating the kinase, rather than through a loss of tumor suppression. By transfecting fibroblast cells with engineered vectors containing different RET mutations in the extracellular domain responsible for MEN 2a and familial MTC (C634Y, C634A, C634W) and MEN 2b (M918T) he was able to demonstrate the increased RET kinase activity and transforming efficiency. MEN 2a mutations altered the disulfide bond between receptor monomers creating an activating homodimer. On the other hand, the MEN 2b mutation altered the substrate specificity for the receptor.[49]

Genetic testing and risk stratification

Since the identification of mutations in the RET gene as the causative agents in MEN 2, genetic tests have been developed and refined to clinically detect these defects.[50-53] D. D. Chi and Wells at Washington University School of Medicine were the first to develop predictive testing for MEN 2 based on genetic mutations in the RET proto-oncogene. They extracted genomic DNA from 96 members of a MEN 2A kindred and PCR-amplified known RET mutations resulting in new restriction endonuclease sites. These new sites could then be detected by digestion with the appropriate enzyme. They then verified the inheritance of the mutation with a previously established genetic linkage test. They found that mutations vary among kindreds but are consistently inherited within kindreds. In addition, they determined that their new direct genetic analysis supplanted an established linkage-based test, since the latter was precluded by recombination events and required the selection of informative genetic markers. The group was able to detect 43 mutations in the 43 affected individuals, thus establishing an invariable correlation between mutation and disease. Importantly, their test identified two genetically affected individuals who were presymptomatic. Thus, the new direct genetic analysis offered a predictive diagnostic test for the disorder.

Today, genetic testing detects nearly 100% of mutation carriers and is considered the standard of care for all first-degree relatives of patients with newly diagnosed MTC. However, due to the varying clinical effects of RET mutations, strategies based on clinical phenotype, age of onset, and aggressiveness of MTC were needed to guide therapy. The first classification system was put forth after the Seventh International Workshop on MEN in 2001, which provided guidelines for the age of genetic testing and prophylactic thyroidectomy.[54] This stratification was revised when the American Thyroid Association (ATA) convened a panel of experts in 2009 to create evidence-based guidelines to assist in the clinical care of MTC patients. Recommendations on the diagnostic workup and timing of prophylactic thyroidectomy and extent of surgery are based on a classification into four risk levels utilizing the genotype-phenotype correlation.[14] The ATA classified mutations according to the risk of developing early, aggressive MTC from A to D in increasing levels. The diagnostic and treatment recommendations based on risk stratification are summarized in Table 1.[55]

Genotype and phenotype

Since the initial discovery of RET mutations responsible for MEN 2, as many as 50 different point mutations across 7 exons (exons 8, 10, 11, 13-16) have been identified.[56] Different mutations in the RET gene produce varying phenotypes for the disease, including age of onset and aggressiveness of MTC, and the presence or absence of other endocrine neoplasms, such as pheochromocytoma or hyperparathyroidism. Approximately 85% of patients with MEN 2 have a mutation of exon 11 codon 634, while mutations in codons 609, 611, 618, and 620 account for 10-15% of cases.[14] Particularly early aggressive behavior and metastasis in MEN 2a and MEN 2b are associated with C634 and M918T mutations, respectively, requiring early intervention.[27] On the other hand, A883F mutation displays a more indolent form of MTC compared with a M918T mutation for MEN 2b.[57] In addition, polymorphism at codon 836 is associated with early metastases in patients with hereditary or sporadic MTC.[58]

Prophylactic thyroidectomy

Untreated patients with MEN 2 ultimately develop MTC and succumb to their disease. Thus, diagnostic tools for detecting MTC in patients at risk for MEN 2 were developed even before the genetic origins were known. Parafollicular cells secrete calcitonin, and blood levels of this hormone serve as a sensitive tumor marker as well as an indicator of the C-cell mass. Intravenously administered calcium and pentagastrin are potent calcitonin secretagogues that markedly enhance the sensitivity of the calcitonin assay. Although the pentagastrin-stimulated calcitonin test provided a sensitive diagnostic test for C cell hyperplasia, it lacked the capacity to predict which patients would develop particularly aggressive forms of MTC and require early intervention and which patients would have a more indolent course and could followed with biochemical surveillance. Because early detection and intervention are paramount to patient survival, genetic tests were needed to prospectively identify MEN 2 patients at-risk for accelerated MTC and requiring immediate treatment.

In 1995, shortly after Wells and others discovered mutations in RET causing MEN 2, Samuel Wells performed thyroidectomies based on identification of these gene mutations in patients. Using PCR-based testing (confirmed by haplotype analysis) for 19 known RET mutations, Wells studied 132 members of seven kindreds with MEN 2a. Twenty-one of the 58 subjects at risk for the disease had germline mutations in the RET oncogene associated with MEN 2a. Plasma calcitonin levels were elevated in 9 of the 21 subjects. However, 12 subjects had normal levels of calcitonin. After undergoing genetic counseling, 13 of the 21 subjects with detected germline mutations in RET, including 6 with normal serum calcitonin levels, underwent total thyroidectomies with central lymph node dissections. All of the surgically resected thyroid glands demonstrated medullary hyperplasia, many with evidence of MTC. However, there was no evidence of lymph node metastasis in any patient at the time of surgery. In addition, postoperative plasma calcitonin levels normalized in patients who had elevated levels preoperatively. Wells concluded that for family members who have germline mutations in the RET proto-oncogene, total thyroidectomy is indicated irrespective of plasma calcitonin levels.[59] Genetic screening and timely thyroidectomy in kindred members who have germline mutated RET alleles characteristic of MEN 2 can prevent MTC, the most common cause of death in these syndromes.[20]

Summary

MEN 2 is an autosomal dominant genetic syndrome caused by missense mutations in the RET proto-oncogene with different penetrance producing three variants, MEN 2a, MEN 2b, and FMTC. Mortality associated with MEN 2 is directly related to the development of MTC for which surgery remains the only chance for cure. As such, recognition of at-risk patients, early diagnosis, genetic risk stratification, timely surveillance, and appropriate surgical intervention are crucial. The initial discovery of the RET proto-oncogene as the causative factor in MEN 2 was the first step in establishing the highly personalized approach to the diagnosis and treatment of MEN 2.

References

1. La Thangue NB, Kerr DJ. Predictive biomarkers: a paradigm shift towards personalized cancer medicine. *Nat Rev Clin Oncol* 2011; 8(10):587-96.
2. Moline J, Eng C. Multiple endocrine neoplasia type 2: An overview. *Genet Med* 2011.
3. Sizemore GW, Health H, 3rd, Carney JA. Multiple endocrine neoplasia type 2. *Clin Endocrinol Metab* 1980; 9(2):299-315.
4. Howe JR, Norton JA, Wells SA, Jr. Prevalence of pheochromocytoma

and hyperparathyroidism in multiple endocrine neoplasia type 2A: results of long-term follow-up. *Surgery* 1993; 114(6):1070-7.

5. Norton JA, Froome LC, Farrell RE, et al. Multiple endocrine neoplasia type IIb: the most aggressive form of medullary thyroid carcinoma. *Surg Clin North Am* 1979; 59(1):109-18.

6. Carney JA, Go VL, Sizemore GW, et al. Alimentary-tract ganglioneuromatosis. A major component of the syndrome of multiple endocrine neoplasia, type 2b. *N Engl J Med* 1976; 295(23):1287-91.

7. Farndon JR, Leight GS, Dilley WG, et al. Familial medullary thyroid carcinoma without associated endocrinopathies: a distinct clinical entity. *Br J Surg* 1986; 73(4):278-81.

8. Sipple J. The association of pheochromocytoma with carcinoma of the thyroid gland. *The American Journal of Medicine* 1961; 31(1):163-166.

9. Underdahl LO, Woolner LB, Black BM. Multiple endocrine adenomas; report of 8 cases in which the parathyroids, pituitary and pancreatic islets were involved. *J Clin Endocrinol Metab* 1953; 13(1):20-47.

10. Manning P, Molnar G, Black B, et al. Pheochromocytoma, hyperparathyroidism, and thyroid carcinoma occurring coincidentally. *N Engl J Med* 1963; 268:68-72.

11. Steiner AL, Goodman AD, Powers SR. Study of a kindred with pheochromocytoma, medullary thyroid carcinoma, hyperparathyroidism and Cushing's disease: multiple endocrine neoplasia, type 2. *Medicine (Baltimore)* 1968; 47(5):371-409.

12. Williams ED, Pollock DJ. Multiple mucosal neuromata with endocrine tumours: a syndrome allied to von Recklinghausen's disease. *J Pathol Bacteriol* 1966; 91(1):71-80.

13. Schimke RN, Hartmann WH, Prout TE, et al. Syndrome of bilateral pheochromocytoma, medullary thyroid carcinoma and multiple neuromas. A possible regulatory defect in the differentiation of chromaffin tissue. *N Engl J Med* 1968; 279(1):1-7.

14. Raue F, Frank-Raue K. Update multiple endocrine neoplasia type 2. *Fam Cancer* 2010; 9(3):449-57.

15. Durbec P, Marcos-Gutierrez CV, Kilkenny C, et al. GDNF signalling through the Ret receptor tyrosine kinase. *Nature* 1996; 381(6585):789-93.

16. Robertson K, Mason I. The GDNF-RET signalling partnership. *Trends Genet* 1997; 13(1):1-3.

17. Nosrat CA, Tomac A, Hoffer BJ, et al. Cellular and developmental patterns of expression of Ret and glial cell line-derived neurotrophic

factor receptor alpha mRNAs. *Exp Brain Res* 1997; 115(3):410-22.

18. Trupp M, Arenas E, Fainzilber M, et al. Functional receptor for GDNF encoded by the c-ret proto-oncogene. *Nature* 1996; 381(6585):785-9.

19. Takahashi M, Ritz J, Cooper GM. Activation of a novel human transforming gene, ret, by DNA rearrangement. *Cell* 1985; 42(2):581-8.

20. Wells SA, Jr., Santoro M. Targeting the RET pathway in thyroid cancer. *Clin Cancer Res* 2009; 15(23):7119-23.

21. Jing S, Wen D, Yu Y, et al. GDNF-induced activation of the ret protein tyrosine kinase is mediated by GDNFR-alpha, a novel receptor for GDNF. *Cell* 1996; 85(7):1113-24.

22. Golden JP, Hoshi M, Nassar MA, et al. RET signaling is required for survival and normal function of nonpeptidergic nociceptors. *J Neurosci* 2010; 30(11):3983-94.

23. Tee JB, Choi Y, Shah MM, et al. Protein kinase A regulates GDNF/RET-dependent but not GDNF/Ret-independent ureteric bud outgrowth from the Wolffian duct. *Dev Biol* 2010; 347(2):337-47.

24. Ryu H, Jeon GS, Cashman NR, et al. Differential expression of c-Ret in motor neurons versus non-neuronal cells is linked to the pathogenesis of ALS. *Lab Invest* 2011; 91(3):342-52.

25. Ohgami N, Ida-Eto M, Sakashita N, et al. Partial impairment of c-Ret at tyrosine 1062 accelerates age-related hearing loss in mice. *Neurobiol Aging* 2011.

26. Pachnis V, Mankoo B, Costantini F. Expression of the c-ret proto-oncogene during mouse embryogenesis. *Development* 1993; 119(4):1005-17.

27. Moore SW, Zaahl MG. Multiple endocrine neoplasia syndromes, children, Hirschsprung's disease and RET. *Pediatr Surg Int* 2008; 24(5):521-30.

28. Reginensi A, Clarkson M, Neirijnck Y, et al. SOX9 controls epithelial branching by activating RET effector genes during kidney development. *Hum Mol Genet* 2011; 20(6):1143-53.

29. Brantley MA, Jr., Jain S, Barr EE, et al. Neurturin-mediated ret activation is required for retinal function. *J Neurosci* 2008; 28(16):4123-35.

30. Schuchardt A, D'Agati V, Larsson-Blomberg L, et al. Defects in the kidney and enteric nervous system of mice lacking the tyrosine kinase receptor Ret. *Nature* 1994; 367(6461):380-3.

31. Schuchardt A, D'Agati V, Larsson-Blomberg L, et al. RET-deficient mice: an animal model for Hirschsprung's disease and renal agenesis. *J Intern Med* 1995; 238(4):327-32.

32. Rungby J. [RET, a gene responsible for familial endocrine neoplasias and Hirschsprung disease]. *Ugeskr Laeger* 1994; 156(21):3194.

33. Lyonnet S, Edery P, Mulligan LM, et al. [Mutations of RET proto-oncogene in Hirschsprung disease]. *C R Acad Sci III* 1994; 317(4):358-62.

34. Edery P, Lyonnet S, Mulligan LM, et al. Mutations of the RET proto-oncogene in Hirschsprung's disease. *Nature* 1994; 367(6461):378-80.

35. Romeo G, Ronchetto P, Luo Y, et al. Point mutations affecting the tyrosine kinase domain of the RET proto-oncogene in Hirschsprung's disease. *Nature* 1994; 367(6461):377-8.

36. Smith DP, Eng C, Ponder BA. Mutations of the RET proto-oncogene in the multiple endocrine neoplasia type 2 syndromes and Hirschsprung disease. *J Cell Sci Suppl* 1994; 18:43-9.

37. Attie T, Edery P, Lyonnet S, et al. [Identification of mutation of RET proto-oncogene in Hirschsprung disease]. *C R Seances Soc Biol Fil* 1994; 188(5-6):499-504.

38. Jeanpierre C, Mace G, Parisot M, et al. RET and GDNF mutations are rare in fetuses with renal agenesis or other severe kidney development defects. *J Med Genet* 2011; 48(7):497-504.

39. Jain S. The many faces of RET dysfunction in kidney. *Organogenesis* 2009; 5(4):177-90.

40. Skinner MA, Safford SD, Reeves JG, et al. Renal aplasia in humans is associated with RET mutations. *Am J Hum Genet* 2008; 82(2):344-51.

41. Gestblom C, Sweetser DA, Doggett B, et al. Sympathoadrenal hyperplasia causes renal malformations in Ret(MEN2B)-transgenic mice. *Am J Pathol* 1999; 155(6):2167-79.

42. Schuchardt A, D'Agati V, Pachnis V, et al. Renal agenesis and hypodysplasia in ret-k- mutant mice result from defects in ureteric bud development. *Development* 1996; 122(6):1919-29.

43. Mathew CG, Chin KS, Easton DF, et al. A linked genetic marker for multiple endocrine neoplasia type 2A on chromosome 10. *Nature* 1987; 328(6130):527-8.

44. Simpson NE, Kidd KK, Goodfellow PJ, et al. Assignment of multiple endocrine neoplasia type 2A to chromosome 10 by linkage. *Nature* 1987; 328(6130):528-30.

45. Santoro M, Rosati R, Grieco M, et al. The ret proto-oncogene is consistently expressed in human pheochromocytomas and thyroid medullary carcinomas. *Oncogene* 1990; 5(10):1595-8.

46. Donis-Keller H, Dou S, Chi D, et al. Mutations in the RET proto-

oncogene are associated with MEN 2A and FMTC. *Hum Mol Genet* 1993; 2(7):851-6.

47. Mulligan LM, Kwok JB, Healey CS, et al. Germ-line mutations of the RET proto-oncogene in multiple endocrine neoplasia type 2A. *Nature* 1993; 363(6428):458-60.

48. Hofstra RM, Landsvater RM, Ceccherini I, et al. A mutation in the RET proto-oncogene associated with multiple endocrine neoplasia type 2B and sporadic medullary thyroid carcinoma. *Nature* 1994; 367(6461):375-6.

49. Santoro M, Carlomagno F, Romano A, et al. Activation of RET as a dominant transforming gene by germline mutations of MEN2A and MEN2B. *Science* 1995; 267(5196):381-3.

50. Unger K, Malisch E, Thomas G, et al. Array CGH demonstrates characteristic aberration signatures in human papillary thyroid carcinomas governed by RET/PTC. *Oncogene* 2008; 27(33):4592-602.

51. Xue F, Yu H, Maurer LH, et al. Germline RET mutations in MEN 2A and FMTC and their detection by simple DNA diagnostic tests. *Hum Mol Genet* 1994; 3(4):635-8.

52. Chi DD, Toshima K, Donis-Keller H, et al. Predictive testing for multiple endocrine neoplasia type 2A (MEN 2A) based on the detection of mutations in the RET protooncogene. *Surgery* 1994; 116(2):124-32; discussion 132-3.

53. Pazaitou-Panayiotou K, Kaprara A, Sarika L, et al. Efficient testing of the RET gene by DHPLC analysis for MEN 2 syndrome in a cohort of patients. *Anticancer Res* 2005; 25(3B):2091-5.

54. Brandi ML, Gagel RF, Angeli A, et al. Guidelines for diagnosis and therapy of MEN type 1 and type 2. *J Clin Endocrinol Metab* 2001; 86(12):5658-71.

55. Kloos RT, Eng C, Evans DB, et al. Medullary thyroid cancer: management guidelines of the American Thyroid Association. *Thyroid* 2009; 19(6):565-612.

56. Frank-Raue K, Rondot S, Schulze E, et al. Change in the spectrum of RET mutations diagnosed between 1994 and 2006. *Clin Lab* 2007; 53(5-6):273-82.

57. Jasim S, Ying AK, Waguespack SG, et al. Multiple endocrine neoplasia type 2B with a RET proto-oncogene A883F mutation displays a more indolent form of medullary thyroid carcinoma compared with a RET M918T mutation. *Thyroid* 2011; 21(2):189-92.

58. Siqueira DR, Romitti M, da Rocha AP, et al. The RET polymorphic allele

S836S is associated with early metastatic disease in patients with hereditary or sporadic medullary thyroid carcinoma. *Endocr Relat Cancer* 2010; 17(4):953-63.

59. Wells SA, Jr., Chi DD, Toshima K, et al. Predictive DNA testing and prophylactic thyroidectomy in patients at risk for multiple endocrine neoplasia type 2A. *Ann Surg* 1994; 220(3):237-47; discussion 247-50.

Table 1. American Thyroid Association Risk Assessment.[55]

ATA risk level	Genetic testing	Neck ultrasound	Serum calcitonin	Thyroidectomy
A	< 3-5 years	> 3-5 years	> 3-5 years	May delay beyond age 5 if normal annual calcitonin and neck US, indolent MTC history, family preference
B	< 3-5 years	> 3-5 years	> 3-5 years	Consider before age 5
C	< 3-5 years	> 3-5 years	> 3-5 years	Before age 5
D	Immediately	Immediately	Immediately	Immediately

MTC = medullary thyroid cancer, PHE = pheochromocytoma, HPT = hyperparathyroidism, HSC = Hirschprung's disease, CLA = cutaneous lichen amyloidosis.

16

The "Miami Criterion" and the Evolution of Minimally Invasive Parathyroidectomy

George L. Irvin III

It is axiomatic that a *failed operation* should initiate the reappraisal of a conventional approach to the management of a common surgical disease. This was nicely exemplified in the development of the technique using the intraoperative measurement of parathormone as a guide to the surgeon performing a parathyroidectomy.

The early history of the conflicting etiologies of hyperparathyroidism and whether the bone disease resulted in or was caused by tumors of the parathyroid gland, the discovery of a hormone secreted by glands which was associated with the disease, and the method to measure parathormone in humans for accurate diagnosis has been well documented by other chapters in this book. Although hyperparathyroidism had been recognized as a surgical disease for more than four decades, the optimum operative approach in the 1960s was still controversial. On one hand, some very prominent surgeons at that time advocated a 3½ gland excision at the initial parathyroidectomy to prevent the high recurrence rate which was reported to be between 15 and 30%.[1-3] On the other hand, several other well-known surgeons reported excellent results by removing only the enlarged, "adenomatous" parathyroid glands, leaving the observed "normal" sized glands intact.[4, 5] Although requiring surgical judgment and experience in deciding which parathyroids were causing the hyperparathyroidism, conservative parathyroidectomy became more widely used when it was recognized that the 3½ gland excision caused tetany in 10% of patients. This complication was often more severe than the original disease and had to be avoided in spite of the occasional persistent disease that followed the less radical operation.

In the next two decades, as the increased use of parathyroid hormone assays evolved, the diagnosis of hyperparathyroidism was made more frequently. Although increased levels of parathormone in patients presenting with hypercalcemia assured a more accurate diagnosis, the necessary extent of the surgical procedure was often unclear. The problem of determining when multiple parathyroid glands were involved and how much tissue to excise remained paramount. It took a long time for many surgeons to realize that hyperparathyroidism had different etiologies and treatments. It was finally recognized that hypersecreting parathyroid glands found in patients with Multiple Endocrine Neoplasia (MEN), or seen after the long term use of certain drugs, and as a consequence of renal failure were different than patients presenting with sporadic primary hyperparathyroidism (SPHPT) where often only one gland was involved. By the late 1980s, many surgeons performed parathyroidectomy on patients with SPHPT using a bilateral neck exploration, with visual identification of all four parathyroid glands, and excision of only the enlarged gland or glands. Using this operative approach, several large centers reported operative success rates of 90 – 95%. However, as the success or failure of parathyroidectomy depended on the experience and judgment of the surgeon, in some hospitals where the procedure was infrequently performed, the operative failure rate was reported to be 30%.[6]

At the University of Miami, Endocrine Surgery was gaining recognition, especially in the area of parathyroid disease in that Dr. Eric Reiss, Chief of Endocrinology and Janet Canterbury, a PhD in his laboratory, had moved there in 1972 and brought with them their famous antibody that had good affinity for human parathyroid hormone. Using a radioimmunoassay, they established the first laboratory capable of providing endocrinologists with a much-needed test that added confirmation to the suspected diagnosis of primary hyperparathyroidism. With their assay, the diagnosis of primary hyperparathyroidism became much more secure, and our surgical referral practice grew as recognition of the usefulness of the parathormone assay began to be appreciated. However, assays available at the time measured the C-terminal, and later the mid-molecule part of the hormone rather than the intact molecule. These fragments had a variable half-life and therefore often yielded clinically confusing results. A major breakthrough came in 1987 when Samuel Nussbaum and his group at the endocrine unit at the Massachusetts General Hospital in Boston described a highly sensitive two-site antibody immunoradiometric assay (IRMA) for measuring intact parathyroid hormone. This assay used two different antibodies, one against the 39-84 segment of the hormone molecule, which was coated on a polystyrene bead, and another antibody against the 1- 34 segment, which was labeled with iodine 125. When these were added together with the unknown amount of intact parathyroid hormone in a sample, very accurate measurements were possible.[7]

In 1988, Nussbaum along with two surgical colleagues presented a report of 13 patients at the American Association of Endocrine Surgeons meeting in Boston suggesting that their PTH IRMA (immunoradiometric assay) technique could be used as an intraoperative adjunct. Their assay had a 15- minute turnaround time and they suggested that this assay could be used as a guide to the extent of neck exploration needed during parathyroidectomy.[8] This report was viewed with interest, but most all of the surgeons in the audience thought that this laboratory technology would be an added burden in the operating room and was not needed when in the hands of an experienced endocrine surgeon hyperparathyroidism could be corrected in 90 to 95% of patients.

At that time, we were performing parathyroidectomies with a thorough bilateral neck and upper mediastinal exploration with attempted visualization of all four parathyroid glands. Excision of the enlarged gland(s) and a safe biopsy of the normal-appearing glands for histological confirmation and DNA analysis was our standard operating procedure. All was well until in February, 1990, when a member of our surgical family came to me with the diagnosis of primary hyperparathyroidism and serum calcium of 13 mg/dL. Rita Martin was special in that she was the operating room supervisor at the University of Miami/Jackson Memorial Hospital and loved by most everyone in the Medical Center. During her parathyroidectomy, we were pleased to find one large parathyroid gland, but despite the usual extensive exploration, only one normal size gland was found and biopsied in the contralateral neck. With the confidence of a 95% success rate we ended the operation at that point. It was extremely embarrassing to me when this patient's calcium failed to fall over the next few days and remained around 12 mg/dL. This failure did not just disappear into the community; everyone in the hospital knew what happened and who the patient's surgeon was! It was obvious that we needed some way of being sure when all abnormal glands had or had not been removed. Could this be done by measuring parathyroid gland hypersecretion during the procedure as suggested by the Boston group two years earlier?

Our research laboratory, supported by the Veterans Administration Hospital, had for some time been involved in flow cytometry and DNA analysis of parathyroid glands trying to establish a method for recognizing the difference between hyperplasia, adenoma, and normal glands. This had had limited success. Immediately after the embarrassing operative failure, we completely changed directions in the laboratory and worked to establish Dr. Nussbaum's previously described methodology to see if we could use it in the operating room. Without buying any equipment, we borrowed a small centrifuge, a gamma counter, a vacuum pump and a few automatic pipettes. With help from the Incstar Corp (Stillwater, MN), which furnished us with some labeled antibodies used in their standard assays, along with a homemade, heated test-tube shaker which was used for enhancement of the reaction time, we assembled all the equipment on a cart. Within 5 months, we had worked out many of the problems for intraoperative use and had an assay that could give us results within 15 minutes. The initial patient for intraoperative PTH analysis at our institution was the same OR supervisor who had an unsuccessful parathyroidectomy five months previously. This is important because no one objected to us bringing this wild-looking machine into the clean operating room if it was going to help us find the bad, overlooked gland causing her disease. At this 2nd procedure a large parathyroid gland hidden in the contralateral thyroid lobe was found and excised with a thyroid lobectomy. The PTH level fell significantly 10 minutes after excision confirming that no other hypersecreting glands were present. She has since remained eucalcemic for more than ten years. This patient was instrumental in convincing us that a quick, intraoperative PTH assay could be of real help to the surgeon by confirming when all hypersecreting glands had or had not been removed during parathyroidectomy.

Over the next several months of study, we measured many serum samples taken at various intervals during every parathyroidectomy and compared them with frozen control PTH levels done later by the standard assay used in the clinical laboratory. The "quick", intraoperative PTH assay (QPTH) had results comparable to the standard assay, but it took some time to coordinate the operative procedure with the best sampling times. We learned early that manipulation of any parathyroid gland by the surgical team could increase the peripheral hormone level and this became a major problem in the education of impatient surgeons as to when to stop exploration in order to measure hormone dynamics accurately. We confirmed that the half-life of PTH was between 3 ½ and 4 minutes, but it was difficult to make the surgeon stop dissecting and searching for other glands after removal of a suspected abnormal parathyroid gland. This was necessary to accurately calculate the hormone degradation and often caused a "tilt" between the laboratory technician team and the surgeon who had not yet learned to trust the intraoperative assay completely, had a busy schedule to keep, and felt it necessary to visualize all four glands before concluding the procedure.

Our early QPTH results were encouraging and we wanted to be the first investigators to prospectively show the clinical advantages of taking a laboratory test into the operating room as in "real-time". In order to quickly get more patients in the study, we were able to solicit help from Dr. Victor Dembrow, a well-known surgeon at Mt. Sinai Medical Center on Miami Beach. To include his patients with hyperparathyroidism in the study, we would load the QPTH cart in an old station wagon for a trip across Biscayne Bay to his operating room. Moving radioisotopes on the cart between the laboratory and the operating rooms in three different hospitals was strictly against the law and Atomic Energy Commission rules, but it was important to have QPTH near the operating surgeon not only to shorten the turn-around time, but to allow good communication between the surgeon and where he or she was in the procedure at the time samples were drawn. When we had enough data for an abstract, it was submitted to the Society of Head and Neck Surgeons where it was accepted for their May, 1991 meeting in Paris. However, due to civil unrest and riots at that time, the meeting site was changed to Maui, Hawaii, and presentation of our data showing what operative monitoring of parathyroid gland hyperfunction offered was well-received, but missed by most of the European investigators interested in this problem.

At the same time, several groups in France were also studying ways to improve their results with parathyroidectomy. Chapuis published the idea of a limited approach with ultrasound localization, local anesthesia, and then confirming that the abnormal gland had been excised first by measuring urinary cAMP and later by plasma hormone levels.[9, 10] These assays took 45 to 80 minutes to run and when more than the excised gland was involved, another operation was necessary with general anesthesia. Around the same time, in 1991, Charles Proye, a very influential surgeon in Lille, France studied the usefulness of intraoperative hormone measurements in his patients and found that the analysis missed some multiple gland involvement; he therefore recommended against the use of this technology.[11] In the same year, Bergenfelz and co-workers at Lund University in Sweden showed clearly that the decline in the plasma level of intact PTH could distinguish between single adenoma and multiglandular disease as the cause of hyperparathyroidism.[12] However, in the United States, we were having a hard time selling this methodology to our surgical colleagues. As we gathered more experience and more confidence in the ability of hormone dynamics to predict when multiple parathyroid glands were involved, we tried to get other surgeons interested in using the test. We even took our famous QPTH cart to the 1992 Miami meeting of the AAES, set it up in a hotel room next to the ballroom and invited as many friends as possible to see how easy it was to measure the hormone level in serum or plasma. Most just smiled and wondered why in the world those crazy surgeons were trying to change their current practice.

Then, a major advancement in technology came when the Nichols Institute Diagnostics made available to us the equipment and labeled antibody for use in an immunochemiluminometric assay in 1993. This new technology did not use radioisotopes, thus making our portable intraoperative assay now legal, and solved the logistical problem of having to schedule elective parathyroid surgery that coincided with the laboratory having enough radioactive labeled-antibody with its short shelf life on hand to perform the assay. This improved method for measuring parathormone had been around since 1987 when Brown and co-workers in Cardiff published their results using a two-site immunochemiluminometric assay (ICMA) with increased sensitivity and specificity.[13] This assay was rapid, stable, and showed no cross reactivity with PTH fragments. Curley and the same group of investigators showed that the ICMA for the intact (1- 84) PTH could distinguish between normal patients, those with hyperparathyroidism, and those with hypercalcemia of malignancy.[14] It was not until six years later we had this improved methodology available in the US.

In 1992 at the University of Miami we began using Tc-99m-sestamibi scans for preoperative localization of abnormal parathyroid glands, and found that in combination with the intraoperative hormone assay, the operative time could be shortened successfully. This new approach to parathyroidectomy was presented in December, 1994 at the Southern Surgical Association meeting in Hot Springs, Virginia. Although there was some real skepticism and pointed questions from the audience, this new and unique methodology gained some national recognition. In the discussion following the presentation, it was shown that with a clearly localized gland, a precise excision could be done, and with intraoperative PTH monitoring assuring a return to eucalcemia, parathyroidectomy could be done on an outpatient basis.[15]

Interestingly, over time, our group found that when parathyroid hypersecretion was used to determine how much tissue to excise, the incidence of multiglandular involvement in primary hyperparathyroidism was much lower than reported by others. While working with us at the University of Miami, Molinari found only 5% of 110 consecutive patients with sporadic primary hyperparathyroidism (SPHPT) in his study had more than one gland involved while others were reporting a frequency varying from 8 to 33%.[16] Only long-term follow-up of recurrence rates would prove if this low incidence of multiple gland involvement predicted by QPTH was justified. Confidence in the ICMA technology as an adjunct to parathyroidectomy was increasing, and Boggs showed that the assay had a sensitivity of 97%, specificity of 100%, and an overall accuracy of 97% for predicting postoperative calcium levels.[17] These two papers were presented at the 1996 AAES meeting in Napa, California and generated some real interest among the surgeons who attended the meeting, especially since the intraoperative assay had become commercially available in the US that year.

As with any new technology, the costs of using the intraoperative assay in the operating room as a surgical adjunct were high, and it was difficult for interested surgeons to convince their hospital administrators of the need for such equipment. The best way to justify this technology was to show that it could not only shorten operating room time, but that parathyroidectomy could be done safely in an outpatient setting without hospital admission. The real cost savings were seen in decreased overnight care. Our friendly local hospital administrators looked at their costs for patients undergoing parathyroidectomy with a same-day discharge and compared them with similar patients undergoing a 1-night hospital stay. Using these cost figures, they were able to come up with a bundled hospital charge for parathyroidectomy that was 39% less if done with a same-day discharge compared to an overnight admission.

We thought that successful outpatient parathyroid surgery would be of great interest to internists, endocrinologists, and primary care physicians, as well as their patients. If a cure could be achieved with minimal discomfort, maybe some more of their symptomatic patients would be sent to the surgeon to benefit from a minimal operative procedure rather than postponing definitive treatment because of the risks and costs of a major operation. To let referring physicians know about this new approach, a report of a consecutive series of 57 patients was submitted for publication to their best medical journal. The article describing the use of intraoperative parathyroid hormone measurement and ambulatory parathyroidectomy was rejected by not one but several prestigious journals, including *The New England Journal of Medicine, Journal of the American Medical Association, Annals of Internal Medicine*, and the *American Journal of Medicine*. Having no success in our attempt to reach this audience of referring physicians, we submitted this report to the American Surgical Association, where it was accepted on the program as an alternate for presentation at the 1996 meeting. When no one died or withdrew an accepted paper for the ASA meeting, we again failed to find an audience. Next, we sent our manuscript to the *Annals of Surgery* for review. When the editorial board also turned it down, despite good reviews, we were really becoming discouraged. This new technology allowing ambulatory parathyroidectomy presented a radical change and none of the editors would take such a risk and publish something so unknown in their journals. Finally, while having breakfast at a meeting with Dr. Claude Organ, I mentioned my frustration and disappointment with the current editorial boards with whom we had been dealing. He asked me face to face if this data and results were "for real?" When I answered with conviction, he encouraged us to resubmit the manuscript to him at the *Archives of Surgery*, where it was finally published a year and a half after the first submission.[18] This report stimulated a great deal of interest in this technology and what it could help accomplish clinically.

In the next three years, three groups presented their results using preoperative SPECT 99m-Tc Sestamibi scans and QPTH to accomplish a more concise, less costly parathyroidectomy at the AAES meetings.[19-21] The group from Johns Hopkins University confirmed our results and showed clearly that minimally invasive parathyroidectomy with local anesthesia could be done safely as outpatient surgery with cost savings compared with patients undergoing bilateral neck exploration under general anesthesia.[21] Others were not convinced that this technology was worthwhile. Several prominent surgeons pointed out in discussion of these presentations that similar results were achieved in their hospitals using the traditional bilateral neck approach and questioned the comparative costs.

To add data showing the benefits of intraoperative parathormone monitoring (IPM), we compared these patients with our previous patients undergoing parathyroidectomy without these surgical adjuncts and demonstrated that the operative failure rates had improved not only in initial operations, but also in patients who had failed operations.[17, 22] Of course not everyone accepted this technology. Over the next few years, several investigators used PTH assays, but with different criteria for predicting postoperative calcium levels. Consequently outcomes were not viewed the same way, and differing results called into question the validity of the methodology. This was emphasized when some surgeons reported their outcomes of parathyroidectomy using different operative indications, definitions of recurrence and operative failure, length of follow-up, and even sometimes, did not separate patients with SPHPT from secondary hyperparathyroidism and MEN. It became obvious that when analyzing the published literature on this subject, the methods used, including how IPM was done, had to be fully explained by each author. Only by using strict definitions could any comparison of the treatment groups be made. In our studies, operative failure is defined as hypercalcemia and elevated PTH level above normal limits within six months after parathyroidectomy. Successful parathyroidectomy is defined as eucalcemia for six months; hypercalcemia and elevated PTH after this time is recurrent hyperparathyroidism. The indications for parathyroidectomy are still controversial and have not been defined clearly in the guidelines published by the AAES. The patients in our studies, symptomatic or not, had biochemical disease with hypercalcemia and elevated PTH above the normal range. Other investigators have included patients with "normocalcemic hyperparathyroidism" and some patients with hypercalcemia and normal PTH levels in their reports. We have not included patients with normocalcemic hyperparathyroidism or hypercalcemia with normal PTH because without the biochemical markers, we believe it is hard to evaluate operative outcome.

The protocols for IPM established by several investigators varied as to site and times of blood sampling in relation to what was going on in the procedure at that time. Each investigative team strived to make the measured hormone dynamics as accurate as possible in predicting complete ablation of all hypersecreting tissue in order to not overlook the patient with multiglandular disease. They all realized that QPTH does well at measuring the hormone level at the specific time when the sample is obtained, and yet used different criteria to predict postoperative calcium levels. The criterion we used called for a greater than 50% drop in PTH from the highest, either preoperative or pre-excision level, ten minutes after the suspected abnormal gland was excised. This became known as the "Miami criterion" and has changed little over the years. We now obtain a 20 minute sample after gland removal if there is a delayed drop in PTH which does not meet or is close to the 50% requirement before continuing with further exploration. Carneiro-Pla published a statistical comparison of several protocols used in IPM that showed the "Miami criterion" to be the most accurate.[23] There were other groups around the country, especially in Atlanta and San Francisco that did not believe that IPM was useful in that they found too many false positive QPTH results causing the hormone assay to miss too many patients with multiglandular disease. This criticism was most evident at the 2008 meeting of the American Surgical Association by the group from the Cleveland Clinic. Siperstein found that IPM failed to identify multiglandular disease in 16% of his patients, risking future recurrence. These investigators would excise a localized abnormal gland, see a significant fall in PTH level, and instead of closing and believing the hormone data which indicated that all hypersecreting glands had been removed, would continue a bilateral exploration and based on size and histopathology find a subsequent gland that was thought to be abnormal.[24] It was pointed out in the discussion of this presentation that the size and histopathology of a parathyroid gland do not correlate with its secretory activity. These investigators used morphology and a diagnosis of "hypercellular" glands from the pathologist as a definition of multiple gland involvement; whereas, surgeons using IPM are guided by only hypersecretion of abnormal glands. Since the success rates of both operative approaches are similar, only the long term follow-up of recurrence rates could prove that the minimally invasive approach, which leaves *in situ* glands of all sizes without hypersecretion, has an outcome as good as the traditional bilateral neck parathyroidectomy. In 2009, using IPM guidance for a focused parathyroidectomy, Lew from the University of Miami reported a recurrence rate of 4% in 164 patients followed seven years, including 43 that were followed for greater than ten years. With 96% of his patients having only one gland removed, he concluded that this operative approach does not fail to identify multiglandular disease and that the various sized glands left in the neck do not cause higher recurrence rates.[25]

Although the developmental history is short, measurement of intraoperative hormone dynamics has broadened our understanding of hyperparathyroidism. It has emphasized that if the hypersecretion of parathormone can be removed with excision of an abnormal gland or glands, long term eucalcemia can be expected. Once it was realized that IPM was capable of assuring the surgeon that all hypersecreting parathyroid glands had been removed, and that it was unnecessary look for other enlarged glands, minimally invasive parathyroidectomy with its many advantages was a logical development. In 2011, "intraoperative parathyroid hormone" has 699 citations on PubMed of the National Library of Medicine and shows that the use of a laboratory test as a guide to the operating surgeon is being implemented all over the world. Thus, real-time measurement of the hormone causing SPHPT has made life easier for endocrine surgeons and their patients.

References

1. Block MA, Greenawald K, Horn RC, et al. Involvement of multiple parathyroids in hyperparathyroidism. Surgical aspects. *Am J Surg* 1967; 114(4):530-7.
2. Haff RC, Black WC, Ballinger WF. Primary hyperparathyroidism:

changing clinical, surgical and pathologic aspects. *Ann Surg* 1970; 171(1):85-92.

3. Paloyan E, Lawrence AM, Oslapas R, et al. Subtotal parathyroidectomy for primary hyperparathyroidism. Long-term results in 292 patients. *Arch Surg* 1983; 118(4):425-31.

4. Attie JN, Wise L, Mir R, et al. The rationale against routine subtotal parathyroidectomy for primary hyperparathyroidism. *Am J Surg* 1978; 136(4):437-44.

5. Wang CA. Surgery of hyperparathyroidism: a conservative approach. *J Surg Oncol* 1981; 16(3):225-8.

6. Malmeus J, Granberg PG, Halvorsen J, et al. Parathyroid surgery in Scandinavia. *Acta Chir Scand.* 1998; 154:405-413.

7. Nussbaum SR, Zahradnik RJ, Lavigne JR, et al. Highly sensitive two-site immunoradiometric assay of parathyrin, and its clinical utility in evaluating patients with hypercalcemia. *Clin Chem* 1987; 33(8):1364-7.

8. Nussbaum SR, Thompson AR, Hutcheson KA, et al. Intraoperative measurement of parathyroid hormone in the surgical management of hyperparathyroidism. *Surgery* 1988; 104(6):1121-7.

9. Chapuis Y, Fulla Y, Icard P, et al. [Peroperative assay of active parathormone 1-84 in surgery of primary hyperparathyroidism]. *Presse Med* 1990; 19(31):1461-2.

10. Chapuis Y, Icard P, Fulla Y, et al. Parathyroid adenomectomy under local anesthesia with intra-operative monitoring of UcAMP and/or 1-84 PTH. *World J Surg* 1992; 16(4):570-5.

11. Proye CA, Goropoulos A, Franz C, et al. Usefulness and limits of quick intraoperative measurements of intact (1-84) parathyroid hormone in the surgical management of hyperparathyroidism: sequential measurements in patients with multiglandular disease. *Surgery* 1991; 110(6):1035-42.

12. Bergenfelz A, Nordén NE, Ahrén B. Intraoperative fall in plasma levels of intact parathyroid hormone after removal of one enlarged parathyroid gland in hyperparathyroid patients. *Eur J Surg* 1991; 157(2):109-12.

13. Brown RC, Aston JP, Weeks I, et al. Circulating intact parathyroid hormone measured by a two-site immunochemiluminometric assay. *J Clin Endocrinol Metab* 1987; 65(3):407-14.

14. Curley IR, Wheeler MH, Aston JP, et al. Studies in patients with hyperparathyroidism using a new two-site immunochemiluminometric assay for circulating intact (1-84) parathyroid hormone. *Surgery* 1987;

102(6):926-31.

15. Irvin GL, Prudhomme DL, Deriso GT, et al. A new approach to parathyroidectomy. *Ann Surg* 1994; 219(5):574-9; discussion 579-81.

16. Molinari AS, Irvin GL, Deriso GT, et al. Incidence of multiglandular disease in primary hyperparathyroidism determined by parathyroid hormone secretion. *Surgery* 1996; 120(6):934-6; discussion 936-7.

17. Boggs JE, Irvin GL, Carneiro DM, et al. The evolution of parathyroidectomy failures. *Surgery* 1999; 126(6):998-1002; discussion 1002-3.

18. Irvin GL, Sfakianakis G, Yeung L, et al. Ambulatory parathyroidectomy for primary hyperparathyroidism. *Arch Surg* 1996; 131(10):1074-8.

19. Carty SE, Worsey J, Virji MA, et al. Concise parathyroidectomy: the impact of preoperative SPECT 99mTc sestamibi scanning and intraoperative quick parathormone assay. *Surgery* 1997; 122(6):1107-14; discussion 1114-6.

20. Inabnet WB, Fulla Y, Richard B, et al. Unilateral neck exploration under local anesthesia: the approach of choice for asymptomatic primary hyperparathyroidism. *Surgery* 1999; 126(6):1004-9; discussion 1009-10.

21. Chen H, Sokoll LJ, Udelsman R. Outpatient minimally invasive parathyroidectomy: a combination of sestamibi-SPECT localization, cervical block anesthesia, and intraoperative parathyroid hormone assay. *Surgery* 1999; 126(6):1016-21; discussion 1021-2.

22. Irvin GL, Molinari AS, Figueroa C, et al. Improved success rate in reoperative parathyroidectomy with intraoperative PTH assay. *Ann Surg* 1999; 229(6):874-8; discussion 878-9.

23. Carneiro DM, Solorzano CC, Nader MC, et al. Comparison of intraoperative iPTH assay (QPTH) criteria in guiding parathyroidectomy: which criterion is the most accurate? *Surgery* 2003; 134(6):973-9; discussion 979-81.

24. Siperstein A, Berber E, Barbosa GF, et al. Predicting the success of limited exploration for primary hyperparathyroidism using ultrasound, sestamibi, and intraoperative parathyroid hormone: analysis of 1158 cases. *Ann Surg* 2008; 248(3):420-8.

25. Lew JI, Irvin GL. Focused parathyroidectomy guided by intra-operative parathormone monitoring does not miss multiglandular disease in patients with sporadic primary hyperparathyroidism: a 10-year outcome. *Surgery* 2009; 146(6):1021-7.

Index